LIST OF TITLES

Already published

A Biochemical Approach to Nutrition	R.A. Freedland, S. Briggs
Biochemical Genetics (second edition)	R.A. Woods
Biological Energy Conservation (second edition)	C.W. Jones
Biomechanics	R.McN. Alexander
Brain Biochemistry (second edition)	H.S. Bachelard
Cellular Degradative Processes	R.T. Dean
Cellular Development	D.R. Garrod
Cellular Recognition	M.F. Greaves
Control of Enzyme Activity	P. Cohen
Cytogenetics of Man and other Animals	A. McDermott
Enzyme Kinetics (second edition)	P.C. Engel
Functions of Biological Membranes	M. Davies
Genetic Engineering: Cloning DNA	D. Glover
Hormone Action	A. Malkinson
Human Evolution	B.A. Wood
Human Genetics	J.H. Edwards
Immunochemistry	M.W. Steward
Insect Biochemistry	H.H. Rees
Isoenzymes	C.C. Rider, C.B. Taylor
Metabolic Regulation	R. Denton, C.I. Pogson
Metals in Biochemistry	P.M. Harrison, R. Hoare
Molecular Virology	T.H. Pennington, D.A. Ritchie
Motility of Living Cells	P. Cappuccinelli
Plant Cytogenetics	D.M. Moore
Polysaccharide Shapes	D.A. Rees
Population Genetics	L.M. Cook
Protein Biosynthesis	A.E. Smith
RNA Biosynthesis	R.H. Burdon
The Selectivity of Drugs	A. Albert
Transport Phenomena in Plants	D.A. Baker
Membrane Biochemistry	E. Sim
Muscle Contraction	C. Bagshaw
Glycoproteins	R.C. Hughes

Editors' Foreword

The student of biological science in his final years as an undergraduate and his first years as a graduate is expected to gain some familiarity with current research at the frontiers of his discipline. New research work is published in a perplexing diversity of publications and is inevitably concerned with the minutiae of the subject. The sheer number of research journals and papers also causes confusion and difficulties of assimilation. Review articles usually presuppose a background knowledge of the field and are inevitably rather restricted in scope. There is thus a need for short but authoritative introductions to those areas of modern biological research which are either not dealt with in standard introductory textbooks or are not dealt with in sufficient detail to enable the student to go on from them to read scholarly reviews with profit. This series of books is designed to satisfy this need. The authors have been asked to produce a brief outline of their subject assuming that their readers will have read and remembered much of a standard introductory textbook of biology. The outline then sets out to provide by building on this basis, the conceptual framework within which modern research work is progressing and aims to give the reader an indication of the problems, both conceptual and practical, which must be overcome if progress is to be maintained. We hope that students will go on to read the more detailed reviews and articles to which reference is made with a greater insight and understanding of how they fit into the overall scheme of modern research effort and may thus be helped to choose where to make their own contribution to this effort. These books are guidebooks, not textbooks. Modern research pays scant regard for the academic divisions into which biological teaching and introductory textbooks must, to a certain extent, be divided. We have thus concentrated in this series on providing guides to those areas which fall between, or which involve, several different academic disciplines. It is here that the gap between the textbook and the research paper is widest and where the need for guidance is greatest. In so doing we hope to have extended or supplemented but not supplanted main texts, and to have given students assistance in seeing how modern biological research is progressing, while at the same time providing a foundation for self help in the achievement of successful examination results.

General Editors:

W.J. Brammar, Professor of Biochemistry, University of Leicester, UK

M. Edidin, Professor of Biology, Johns Hopkins University, Baltimore, USA

Control of Enzyme Activity

Philip Cohen

Professor, Department of Biochemistry
University of Dundee

Second Edition

Chapman and Hall
London and New York

First published 1976
by Chapman and Hall Ltd
11 New Fetter Lane, London EC4P 4EE
Second Edition 1983
Published in the USA
by Chapman and Hall
733 Third Avenue, New York NY 10017

Printed in Great Britain by
J. W. Arrowsmith Ltd, Bristol

ISBN 0 412 25560 X

British Library Cataloguing in Publication Data

Cohen, Philip
    Control of enzyme activity.—2nd ed.—(Outline
    studies in biology)
    1. Enzymes
    I. Title   II. Series
    574.19'25   QP601
ISBN 0-412-25560-X

Library of Congress Cataloging in Publication Data

Cohen, P. (Philip), 1945–
    Control of enzyme activity.
    (Outline studies in biology)
    Bibliography: p.
    Includes index.
    1. Enzymes—Addresses, essays, lectures.
2. Metabolism—Regulation—Addresses, essays, lectures.
I. Title. II. Series: Outline studies in biology
(Chapman and Hall)   [DNLM: 1. Enzymes—Metabolism.
QU 135 C676c]
QP601.C59 1983   574.1'925   83-7538
ISBN 0-412-25560-X

Contents

This book is dedicated to Professor Carl Cori, still very active in science in his 87th year.

Acknowledgement

I am greatly indebted to Dr Michael Kerr for invaluable assistance in the revision of Chapter 3, and to Drs Earl Davie, Earl Stadtman, Richard Kahn, Alfred Gilman and Louise Johnson for providing information in advance of publication.

1 Introduction

'A principal goal of the study of regulation of enzyme activities in cells or tissues is directed towards an understanding of the mechanisms which link metabolism to function' [1]

The metabolic charts which adorn the walls of biochemical laboratories give some impression of the vast number of enzyme catalysed reactions which take place in living cells. They are also a reminder of the magnitude of the problem that confronts an organism in harnessing the various pathways, so that they deliver metabolites to the right place, in the right amount and at the right time, and at a minimum cost in terms of energy. It has become clear, that a major way in which such regulation is achieved is by endowing certain enzymes stationed at key points in metabolism, with special properties.

It is self evident, that the rate at which the ultimate products of a metabolic pathway are formed, can only be controlled by changing the activity of the *rate limiting* enzyme in that pathway. This can be achieved by either increasing and decreasing the number of enzyme molecules (induction and repression), or by changing the activity of pre-existing enzyme molecules. This book is concerned solely with the latter type of effect.

The study of enzyme regulation has two major goals.

(a) To identify the enzyme in a pathway which is rate limiting at a particular state of metabolic activity.

(b) To characterize the mechanisms which regulate the activity of the rate limiting enzyme *in vivo*, and to understand how such controls relate to the integrated metabolism of the cell.

It is the second of these goals that is the main concern of this study. Another volume of this series describes the methods which are used to identify rate limiting steps in greater detail [2].

By the very nature of the subject, the control of enzyme activity encompasses the whole of Biochemistry. The book must therefore pre-suppose that the reader has a broad familiarity with the pathways of metabolism, how they interrelate with one another, and some understanding of basic concepts in protein chemistry and enzyme kinetics.

The enzyme systems which are used to discuss this topic are drawn from a wide range of biological systems. An attempt has been made to describe these in sufficient depth, that both the limitations of currently available techniques and the way in which future research is likely to proceed will be appreciated. Perhaps the most striking feature to emerge from a first reading, is the great variety of mechanisms which organisms have selected for the control of enzyme activity. Hopefully, some more general themes will start to emerge as the book progresses, and the brief summaries at the end of each chapter are designed to help the reader reach an overall synthesis of the 'current state of the art'.

References

[1] Helmreich, E. and Cori, C. F. (1966), *Adv. Enz. Reg.*, **3**, 91−107.
[2] Denton, R. M. and Pogson, C. I. (1976), *Metabolic Regulation*, Chapman and Hall, London.

2 Regulation of amino acid and nucleotide biosynthesis in bacteria by end product inhibition

Given minimal growth media, containing carbon, hydrogen, nitrogen, oxygen and sulphur, bacteria such as *E. coli* can synthesize an enormous variety of metabolites, including all the 20 amino acids required to manufacture proteins, and each of the nucleotides for RNA and DNA production. In contrast, higher organisms, such as mammals, lack many of the enzymes which catalyse these reactions, and these compounds are often essential nutrients in the diet. The extra versatility of bacteria can be thought of as being counterbalanced by the need to expend large amounts of carbon, nitrogen and energy to synthesize not only every amino acid and nucleotide, but also all the enzymes required for these biosynthetic reactions (numbering over 100 for amino acid biosynthesis alone). It is therefore not surprising, that by the mid 1950s, when the metabolic routes involved had been largely worked out, that attention had already centred on how these reactions might be regulated to make the most effective use of available nutrients. Indeed, many of the isotopic and mutant studies which had been used to elucidate the sequence of steps in these pathways had themselves given the first clues as to how these controls might operate.

In one such series of experiments [1], a culture of *E. coli*, growing exponentially on a glucose–salt medium, was centrifuged and resuspended in the same medium, except for the replacement of the unlabelled glucose by radioactive ^{14}C-glucose. The incubation was continued for a further hour, by which time the number of bacteria had approximately doubled. Under these conditions of logarithmic growth, bacteria are actively synthesizing protein, and contain the full complement of enzymes for the synthesis of every amino acid. If the bacterial protein was then analysed, the radioactivity associated with each amino acid approximated, as expected, to its relative content in the total bacterial protein. If, however, the growth medium was supplemented with an unlabelled amino acid, such as L-isoleucine, radioactivity incorporated into protein isoleucyl residues fell by more than 95%, and analogous observations were made when other amino acids were used. The experiments showed that the added amino acid was being used preferentially, and that in some way, the amino acid was inhibiting its synthesis from precursor molecules.

2.1 The control of L-isoleucine biosynthesis
This metabolic pathway is shown in Fig. 2.1, which also gives the synthetic route for the structurally related amino acid, L-valine. The experiment described above had shown that the presence of L-isoleucine in the growth medium stopped the intracellular formation of L-isoleucine. Furthermore, L-isoleucine added to the growth medium of a bacterial mutant which

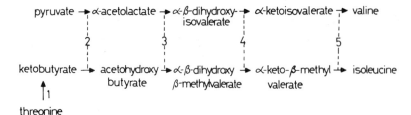

Fig. 2.1 Biosynthetic pathways for the formation of L-isoleucine and L-valine. Enzymes: 1–threonine deaminase; 2–acetohydroxyacid synthetase; 3–acetohydroxy-acid isomeroreductase; 4–dihydroxyacid dehydrase; 5–transaminase B and, for the valine pathway only, valine-alanine aminobutyrate transaminase [2].

could not by itself make L-threonine, reduced the amount of L-threonine required to achieve an optimal growth rate [3]. This indicated that not only was L-threonine required for L-isoleucine biosynthesis, but that L-isoleucine had in some way decreased the activity of one or more of the enzymes which converted L-threonine to L-isoleucine. Since the first step in this sequence, catalysed by threonine deaminase (Fig. 2.1) is irreversible, L-isoleucine could not have exerted its effect by reversing the flow of the pathway from L-isoleucine back to L-threonine.

The discovery that threonine deaminase was inhibited by L-isoleucine *in vitro* [4] was an important landmark in the understanding of enzyme control mechanisms. The inhibition was very specific, in that L-leucine was 100 times less effective, while L-valine and D-isoleucine had no inhibitory effect whatsoever. Moreover, no other enzyme in the pathway was inhibited by L-isoleucine *in vitro*. This immediately suggested a very simple and attractive mechanism, requiring no extra expenditure of energy, which could explain how L-isoleucine and other amino acids were able to regulate their own biosynthesis. If this mechanism occurred in the intact cell, increasing concentrations of L-isoleucine would inhibit threonine deaminase, the first enzyme in the pathway, and decrease new L-isoleucine biosynthesis by suppressing the metabolic flow through the entire pathway. Conversely, if the levels of L-isoleucine fell, its production from L-threonine would increase automatically, since the effect was completely reversible.

The inhibition of threonine deaminase by L-isoleucine was found to be competitive, in that the concentration of L-threonine required for half maximal activation was increased, but the kinetics were complex. If the initial rate of L-threonine deamination was plotted against L-threonine concentration in the absence of L-isoleucine, an S-shaped or *sigmoid* curve was obtained and not a simple hyperbolic function [5]. An analogous result was also obtained if threonine deamination was plotted against L-isoleucine at a fixed concentration of L-threonine (Fig. 2.2). L-isoleucine had to be increased above a certain *threshold* level, before marked inhibition occurred. How such unusual kinetic behaviour arises is considered in Chapter 6, but for the present, it is sufficient to note that the shape of the curve renders the activity more sensitive to end product inhibition over a particular range of L-isoleucine than would be the case if the enzyme merely

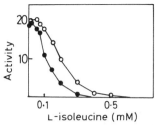

Fig. 2.2 Effect of L-isoleucine on L-threonine deaminase activity in a crude extract of *E. coli* K12. The assays contained 0.02 M (●) or 0.04 M (○) L-threonine [5].

exhibited simple Michaelis Menten kinetics. This suggested that the complex kinetics of threonine deaminase observed *in vitro* might be physiologically important, in that they allowed a more sensitive regulation of the pathway in response to fluctuations in the L-threonine and L-isoleucine concentration.

2.2 Desensitization of threonine deaminase to L-isoleucine: the allosteric theory

When threonine deaminase was heated, it was found to lose its sensitivity to inhibition by L-isoleucine, although the activity measured in the absence of L-isoleucine remained unaffected (Fig. 2.3). This desensitization effect was an important finding, because it implied that L-isoleucine and L-threonine must bind to different sites on the enzyme. The effect of L-isoleucine on the binding of L-threonine could therefore only be indirect.

In a classic review, Monod, Changeux and Jacob [6] describing the then known properties of threonine deaminase and other enzymes subject to metabolic control, introduced the term *allosteric effector* to describe a regulatory molecule such as L-isoleucine, which inhibited (or activated) a particular enzyme. The whole concept is beautifully described in the following paragraph taken from their paper. 'The allosteric effector binds specifically and reversibly to the *allosteric site*. The formation of the enzyme allosteric effector complex does not activate a reaction involving the effector itself, but is assumed to bring about a discrete reversible alteration in the molecular structure of the protein, the *allosteric transition*, which modifies the properties of the active site, changing one or several of the kinetic parameters which characterize the biological activity of the protein. An absolutely essential, albeit negative assumption implicit in this

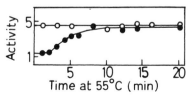

Fig. 2.3 Desensitization of threonine deaminase by heat. Activity was measured with 0.02 M L-threonine in the presence (●) or absence (○) of 0.01 M L-isoleucine [5].

description is that an allosteric effector, since it binds at a site altogether distinct from the active site, and since it does not participate at any stage of the reaction activated by the protein, need not bear any particular chemical or metabolic relation of any sort with the substrate itself. The specificity of any allosteric effect and its actual manifestation is therefore considered to result exclusively from the specific construction of the protein molecule itself, allowing it to undergo a particular discrete, reversible conformational alteration, triggered by the binding of the allosteric effector. The absence of any inherent obligatory chemical analogy or reactivity between substrate and allosteric effector appears to be a fact of extreme biological importance'.

2.3 Regulation of L-threonine deaminase activity *in vivo*

The specific inhibition of L-threonine deaminase by L-isoleucine observed *in vitro* suggested, but did not prove, that this was how L-isoleucine prevented further L-isoleucine biosynthesis when it was added to the growth medium. The most convincing demonstration of the importance of this effect *in vivo* came from studying the behaviour of bacterial mutants which synthesized an altered threonine deaminase, desensitized to inhibition by L-isoleucine. These bacteria overproduced L-isoleucine to such an extent, that this amino acid was excreted into the culture medium in large quantities [7].

Although L-valine does not inhibit threonine deaminase, it was subsequently shown to *activate* the enzyme [7,8] at low L-threonine concentrations by abolishing the sigmoidal dependence of activity on L-threonine concentration (Fig. 2.4). The possible role of this effect can be illustrated by reference to Fig. 2.1. This shows that the pathways of L-isoleucine and L-valine biosynthesis are very closely interrelated in that they share three enzymes. Accordingly, activation by L-valine might increase L-isoleucine biosynthesis under conditions where L-threonine is limiting, and thus help to balance L-valine and L-isoleucine biosynthesis. However, whether such a stimulation of a *parallel pathway* occurs *in vivo* has not yet been established.

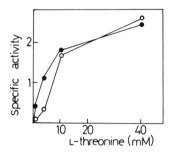

Fig. 2.4 Activity of threonine deaminase plotted as a function of threonine in the presence (●) or absence (○) of 5.0 mM L-valine [8].

2.4 The control of L-lysine, L-methionine, L-threonine and L-isoleucine biosynthesis

The biosynthesis of L-isoleucine from L-threonine is a relatively simple pathway, as is apparent from an inspection of Fig. 2.5. Although it has been considered up till now as an independent metabolic route, in effect, it is only a branch of a much wider pathway, which culminates in the synthesis of four different amino acids, L-lysine, L-methionine, L-threonine and L-isoleucine, from the common precursor, L-aspartic acid. Such a situation immediately poses more complex problems of metabolic control. For example, if an early enzyme in the pathway, such as aspartokinase, was efficiently inhibited by just one of the four amino acid end products, the supply of intermediates for the synthesis of the other three essential metabolites would also become depleted, creating severe problems for the bacterium. The structure of the pathway itself therefore suggests that a more sophisticated control mechanism is likely to be necessary.

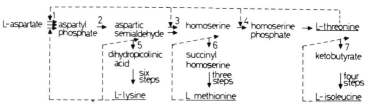

Fig. 2.5 The biosynthesis of L-lysine, L-methionine, L-threonine and L-isoleucine from L-aspartic acid. End product inhibitions are indicated by the broken lines. Multiple arrows denote isoenzymes. 1—aspartokinase; 2—aspartic semialdehyde dehydrogenase; 3—homoserine dehydrogenase; 4—homoserine kinase; 5—dihydropicolinic acid synthetase; 6—succinyl homoserine synthetase, 7-threonine deaminase.

2.5 Isoenzymes of aspartokinase in *E. coli* K12, with different regulatory properties

A similar experiment to that described earlier for threonine deaminase, revealed that L-lysine and L-threonine were the only two amino acids that inhibited aspartokinase activity in bacterial extracts. Inhibition by either amino acid never exceeded 40–50%, but the effects were additive, so that inhibition was almost complete when the two amino acids were both present in excess (Fig. 2.6). This unexpected phenomenon was explained, when aspartokinase activity was fractionated into two components by standard protein purification techniques. The first component could be completely inhibited by L-thronine but was unaffected by L-lysine, while the second component could be completely inhibited by L-lysine but was unaffected by L-threonine [9].

In the presence of excess L-lysine, aspartyl phosphate production should therefore be reduced but not diminished to a point where none is available for L-threonine biosynthesis and vice versa. However, for these controls to be effective, additional devices are clearly necessary to ensure that in the presence of excess L-lysine, the reduced amount of aspartyl phosphate is channelled towards L-threonine production, and that in the presence of

13

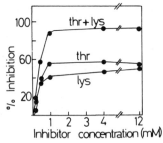

Fig. 2.6 Effect of L-threonine and L-lysine on aspartokinase activity in extracts of *E. coli* K12 [9].

excess L-threonine, the reduced aspartyl phosphate is channelled towards L-lysine. This led to the finding that the first enzyme on the *branch* leading to L-lysine, dihydropicolinic acid synthetase, was specifically inhibited by L-lysine [10], while the first enzyme on the branch leading to L-threonine, homoserine dehydrogenase, was specifically inhibited by L-threonine [11]. These two secondary controls should ensure that the reduced aspartyl phosphate is shunted over to either L-lysine or L-threonine when one of these amino acids is present in excess (Fig. 2.5).

However, homoserine is an intermediate for L-methionine as well as L-threonine biosynthesis (Fig. 2.5). If there was just a single homoserine dehydrogenase which was completely inhibited when L-threonine was in excess, L-methionine synthesis would also be suppressed completely, unless yet a further control mechanism was present. This led to the discovery that there was a third aspartokinase and a second homoserine dehydrogenase in *E. coli* K12. These two activities were not inhibited by L-lysine, L-threonine or L-methionine, although their synthesis was strongly repressed by L-methionine. Indeed, it was necessary to work under conditions of L-methionine limitation with mutants which could not themselves make L-methionine before these activities were derepressed sufficiently to be easily detectable [12]. These two activities should enable L-menthionine biosynthesis to continue when both L-lysine and L-threonine are present in excess.

However, the presence of two homoserine dehydrogenases then implies that a further mechanism is needed to ensure that in the presence of excess L-threonine, the reduced homoserine is channelled towards L-methionine and not L-threonine production, and vice versa. This is probably achieved through the inhibition of homoserine kinase by L-threonine [13] and the inhibition of O-succinyl homoserine synthetase by L-methionine [14], which have been demonstrated *in vitro*.

The importance of some, but not all of these controls *in vivo* has been established by using bacterial mutants in a manner analogous to that described earlier for threonine deaminase (Section 2.3).

The biosynthesis of the three aromatic amino acids from the common precursor DAHP in *E. coli* appears to be regulated in a perfectly analogous manner (Fig. 2.7). There are three DAHP synthetases, one inhibited by

14

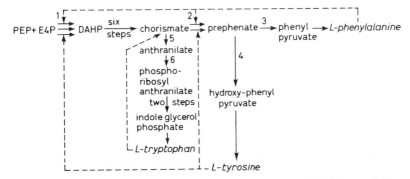

Fig. 2.7 Biosynthesis of the aromatic amino acids. End product inhibitions are indicated by the broken lines. Multiple arrows denote isoenzymes. 1–3, deoxy-D-arabino-heptulosonic acid-7-phosphate (DAHP) synthetase; 2–chorismate mutase; 3–prephenate dehydratase; 4–prephenate dehydrogenase; 5–anthranilate synthetase; 6–anthranilate-5-phosphoribosyl pyrophosphate (PRPP) phosphoribosyl transferase. PEP–phosphoenol pyruvate; E4P–erythrose-4-phosphate.

L-phenylalanine, one by L-tyrosine, and a third which is not subject to feedback inhibition but whose synthesis is repressed by L-tryptophan [15]. Later in the sequence after the tryptophan pathway diverges, there are two chorismate mutases, one inhibited by L-phenylalanine (chorismate mutase P) and the other by L-tyrosine (chorismate mutase T) [16]. These examples show that the use of two or more enzymes catalysing the same biochemical reaction (termed *isoenzymes*) is a common device for regulating branched metabolic pathways in bacteria.

2.6 Bifunctional aspartokinase-homoserine dehydrogenases: multifunctional enzymes and multienzyme complexes

When bacterial mutants which excreted large quantities of L-threonine into the growth medium, were analysed, the unexpected finding was made, that threonine sensitive aspartokinase and threonine sensitive homoserine dehydrogenase were both partially or completely desensitized to inhibition by L-threonine [11]. Since a single mutational event had affected two activities in an identical manner, this strongly suggested that the two enzymes must at least share a common polypeptide chain. This idea was reinforced by kinetic studies, which showed that L-aspartate and ATP, the substrates for aspartokinase inhibited homoserine dehydrogenase activity *in vitro*, while NADP and homoserine, the products of the homoserine dehydrogenase reaction, inhibited aspartokinase. Furthermore, NADPH, a substrate of homoserine dehydrogenase protected threonine sensitive but *not* lysine sensitive aspartokinase against thermal denaturation. Conclusive proof that the two activities were carried by the same protein was obtained when the two activities were shown to copurify through a number of purification steps, and the final product was found to be a homogenous protein, termed aspartokinase I–homoserine dehydrogenase I. The protein (M_r 350 000) was shown to be composed of four identical subunits held together by non-covalent forces [17,18]. Consequently, the two catalytic activities must be located on a single polypeptide chain.

A similar analysis showed that the aspartokinase and homoserine dehydrogenase activities which were repressed by growth on L-methionine were also carried by a single protein, termed aspartokinase II—homoserine dehydrogenase II. The lysine sensitive aspartokinase is referred to as asparto-kinase III [11].

When aspartokinase I—homoserine dehydrogenase I was incubated with chymotrypsin, aspartokinase activity was destroyed and homoserine dehydrogenase activity became desensitized to inhibition by L-threonine. A fragment (M_r 55 000) was isolated, which contained the desensitized homoserine dehydrogenase activity, and was shown by endgroup analysis to correspond to the C-terminal portion of the polypeptide chain. A further mutant of aspartokinase I — homoserine dehydrogenase I was then obtained, that had normal aspartokinase activity sensitive to inhibition by L-threonine, but was devoid of homoserine dehydrogenase activity. The mutant protein was isolated (M_r 47 000) and shown to correspond to the amino-terminal portion of the chain. The mutation is of the 'Ochre' type, premature chain termination being responsible for the shortened polypeptide chain[19]. The isolation and analysis of these fragments proved that the two activities are carried by *distinct* regions of a single polypeptide chain (Fig. 2.8). Since the two fragments retain one or other of the two activities with unchanged substrate affinities, the polypeptide must be folded into two distinct globular domains, each region retaining its own independent activity when separated from the other.

Many *bifunctional* enzymes have now been identified in which two activities are carried on the same polypeptide. These include the first two enzymes of tryptophan biosynthesis in *E. coli* (Fig. 2.7), anthranilate synthetase and anthranilate-PRPP phosphoribosyl transferase [20] and further examples can be found in Chapters 4 and 5. Indeed a number of *multifunctional* enzymes have been recognised in which *more* than two activities reside in the same polypeptide. This situation appears to be much more common in eukaryotic than prokaryotic organisms. For example, in mammalian cells the first three enzymes of pyrimidine biosynthesis, carbamyl phosphate synthetase, aspartate transcarbamylase and dihydro-orotase (Fig. 2.9) are associated with a single M_r 215 000 polypeptide [21], while in the common bread mould *Neurospora crassa*, five consecutive

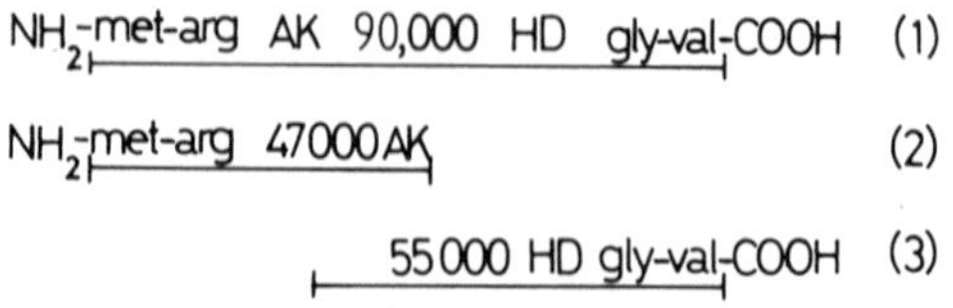

$$NH_2\text{-met-arg} \quad AK \quad 90,000 \quad HD \quad gly\text{-val-}COOH \quad (1)$$

$$NH_2\text{-met-arg} \quad 47000\,AK \quad (2)$$

$$55000 \quad HD \quad gly\text{-val-}COOH \quad (3)$$

Fig. 2.8 The structure of aspartokinase I-homoserine dehydrogenase I from *E. coli* K12. 1—native enzyme; 2—N-terminal fragment from 'Ochre' mutant possessing aspartokinase (AK) activity; 3—C-terminal fragment from a chymotryptic digest possessing homoserine dehydrogenase (HD) activity.

16

enzymes of aromatic amino acid synthesis that convert DAHP into 5-enoylpyruvoylshikimate (Fig. 2.7) are located on an M_r 165 000 polypeptide [22]. In mammalian cells the seven enzymes required to synthesize long chain saturated fatty acids from malonyl CoA are all contained within an M_r 250 000 protein [23]. In contrast, all these enzymes of aromatic amino acid, pyrimidine and fatty acid synthesis exist as separate entities in *E. coli*. The synthesis of the cyclic peptide antibiotic *enniaten* in the plant fungus *Fusarium oxysporum* involves at least five enzymes that are all located on an M_r 250 000 polypeptide [24].

One possible way in which such enzymes may have evolved is by gene fusion. This has been carried out experimentally in *S. typhimurium*, by fusing the second and third genes of the histidine operon, which normally code for two separate proteins, histidinol dehydrogenase and imidazole-acetol-phosphate-amino-transferase. The fusion was achieved by inducing two sequential frameshift mutations near the intercistronic region, which eliminated the chain termination codon normally present between the two genes. As a result, the two enzymes were synthesized as a single polypeptide which folded in such a way as to retain a substantial proportion of both catalytic activities [25].

In multifunctional enzymes, or multienzyme complexes where several enzymes are linked to one another by non-covalent forces, the various activities are closely related functionally, and indeed often catalyse a series of directly connected reactions. There are two obvious advantages of such a situation:

(a) The sequence of reactions become localized in a particular region of the cell, so that the product formed by one reaction becomes available to the next enzyme of the sequence at a much higher concentration than would be the case if the enzymes were dispersed randomly throughout the cell.

(b) Since the product of one reaction can be channelled directly to the next enzyme and not released from the surface of the complex, unwanted side reactions that the product might otherwise undergo are minimized. The chorismate mutase system in aromatic amino acid biosynthesis (Fig. 2.7) illustrates this rather nicely. Chorismate mutase-P (inhibited by phenylalanine) is complexed with prephenate hydratase, the enzyme that forms phenylalanine, while chorismate mutase-T (inhibited by tyrosine) is complexed with prephenate dehydrogenase, the enzyme that forms tyrosine [16]. This presumably results in a situation where there are effectively two pools of prephenate, one specifically channelled to phenylalanine and the other to tyrosine production.

Unfortunately, the situation is more difficult to understand in the case of the two bifunctional aspartokinase-homoserine dehydrogenases, because they are the *first* and *third* enzymes of the pathway and do not follow one another (Fig. 2.5). There is no evidence that the second enzyme, aspartic semialdehyde dehydrogenase (of which there appears to be just one type) is linked to these bifunctional enzymes [11]. In the case of aspartokinase I—homoserine dehydrogenase I, it could be argued that if the gene for a separate aspartokinase sensitive to inhibition by L-threonine became fused to a gene coding for homoserine dehydrogenase, then very few mutational events might be needed to create interaction between the two globular domains, resulting in a structure in which L-threonine coordinately inhibited homoserine dehydrogenase as well as aspartokinase. The economy of such a situation might then have a selective advantage [18]. However, this argument cannot be used for aspartokinase II–homoserine dehydrogenase II which is not subject to feedback control. However, in all bifunctional enzymes, the two activities must be synthesized in equimolar quantities; the linkage is therefore a simple means of coordinating the synthesis of both activities.

2.7 Regulation of aspartokinase in different bacteria: the evolution of control mechanisms

Aspartokinase I–homoserine dehydrogenase I is only carried by a single polypeptide chain in *E. coli* and *Salmonella typhimurium*. In other bacteria, such as *Azotobacter* and *Pseudomonas*, the two enzymes are made as two separate proteins *each* inhibited by L-threonine. In *B. polymyxa* and *B. subtilis*, there appears to be just one aspartokinase, which is not inhibited by either L-threonine or L-lysine, but both L-threonine and L-lysine *together* produce an almost total inhibition of activity (multivalent feedback inhibition). In *Rhodospirillum rubrum*, there appears to be a single aspartokinase which can be completely inhibited by L-threonine. However, L-methionine and L-isoleucine can activate this enzyme in the absence of L-threonine, while L-isoleucine can effectively reverse the inhibition by L-threonine. The threonine deaminase of this bacterium is not sensitive to inhibition by L-isoleucine. Since there appears to be no effect of L-lysine on aspartokinase activity in this organism, an increase in the isoleucine/threonine ratio appears to be the signal for accelerated production of common intermediates required for L-lysine and L-methionine biosynthesis [26,27].

Although the enzymes of these other bacteria have not been characterized to anything like the extent of the *E. coli* K12 system, it is clear that different bacteria have solved the problem of the regulation of a given pathway in quite different ways. It is an extremely important general rule, that although the steps in a metabolic pathway are essentially invariant between organisms, the regulation of these pathways can vary significantly, not only from organism to organism, but in the case of mammals, from cell type to cell type within a given animal. The aspartokinase system illustrates the fascinating area of the development of control mechanisms, which undoubtedly evolved long after the pathways themselves.

2.8 The control of pyrimidine biosynthesis in *E. coli*

The biosynthesis of UTP and CTP is another metabolic pathway which has been important in the development of ideas about the control of enzyme activity. The steps in the pathway and the major allosteric effects are summarized in Fig. 2.9.

One of the original experiments [28] that implicated aspartate trans-

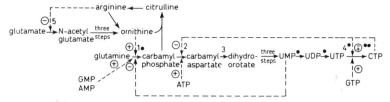

Fig. 2.9 Regulation of pyrimidine biosynthesis in *E. coli* (+), allosteric activator; (−), allosteric inhibitor; •−reactions using ATP as a substrate; 1−carbamyl phosphate synthetase; 2−aspartate transcarbamylase; 3−dihydroorotase; 4−CTP synthetase; 5−N-acetylglutamate synthetase. CTP can be either an activator or inhibitor of CTP synthetase (see text).

carbamylase (ATCase) in the regulation of the pathway is shown in Fig. 2.10. In this study, an *E. coli* mutant was used which lacked dihydroorotase, the enzyme that follows ATCase (Fig. 2.9). As a result, the mutant could catalyse the ATCase reaction but still required uracil for growth. Just prior to the time at which uracil in the growth medium would be exhausted, a mutant culture was divided into two portions, and one portion was supplemented with excess uracil. In the other portion, growth soon ceased due to uracil starvation, and carbamyl aspartate started to accumulate on a massive scale reaching 50% of the dry weight of the bacterium within four hours. Addition of uracil abolished further carbamyl aspartate production very rapidly, while the portion given uracil before the time when uracil starvation would have occurred never accumulated this compound. This showed that ATCase activity was inhibited *in vivo* by uracil or a metabolite derived from it, and this led to the discovery that CTP was a potent feedback inhibitor of the enzyme *in vitro*. CMP only inhibited weakly, while

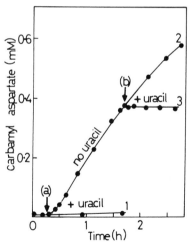

Fig. 2.10 Effect of uracil starvation on carbamyl aspartate production in an *E. coli* mutant lacking dihydroorotase. (a) point of uracil exhaustion. Just prior to this time, a portion of the culture (1) was supplemented with 0.4 mM uracil; (b) at this point a further portion (3) was supplemented with 0.4 mM uracil [28].

UTP and UMP had hardly any effect at all [28,29]. The importance of CTP inhibition *in vivo*, has been demonstrated by using a mutant lacking CTP synthetase (Fig. 2.9). This mutant could make UTP by the pyrimidine pathway, but required cytidine added exogenously in order to make CTP. When cytidine was withheld, intracellular UTP rose 10-fold and uridine was excreted into the growth medium, indicating that despite the presence of large amounts of UTP, the pyrimidine pathway was not being controlled. In the presence of cytidine, and hence CTP, over-production of UTP did not occur [30,31].

ATP is an activator of ATCase *in vitro* [28,29]. If ATCase activity is plotted as a function of aspartate, a sigmoid curve is obtained. CTP competes with aspartate, shifting the sigmoid curve to the right and making it more sigmoidal (Fig. 2.11). ATP shifts the curve to the left making it less sigmoidal, and also antagonizes the inhibition by CTP. Since the four nucleoside triphosphates ATP, GTP, CTP and UTP must be present in approximately equal amounts in order not to limit RNA synthesis, activation of ATCase may help to stimulate UTP and CTP synthesis when L-aspartate is low, thus balancing purine and pyrimidine biosynthesis. The effects of ATP and CTP on ATCase activity are analogous to those of L-valine and L-isoleucine on threonine deaminase activity (Figs 2.2, 2.4)

Further mechanisms for balancing UTP, CTP, ATP and GTP production may occur at the level of CTP synthetase (Fig. 2.9). The activity of this enzyme requires the presence of the allosteric activator GTP [32]. CTP inhibits CTP synthetase when its concentration is high, probably by competing with the substrate UTP, but activates at low UTP concentrations, probably by replacing GTP. ATP is also a substrate of CTP synthetase as well as three further enzymes of the pyrimidine pathway (Fig. 2.9).

The enzyme that precedes ATCase, carbamyl phosphate synthetase (Fig. 2.9) is also subject to control by both purine and pyrimidine nucleotides [33–35]. It is strongly inhibited by UMP, and also UDP and UTP,

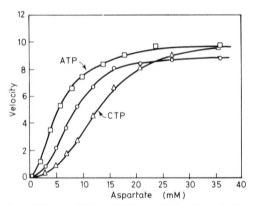

Fig. 2.11 Effect of ATP and CTP on ATCase activity. The velocity was measured at pH 7.0 in the absence of allosteric effectors (○), or in the presence of 0.4 mM CTP (△) or 2.0 mM ATP (□) [39].

20

and activated by purine nucleotides such as IMP, GMP and AMP. The other use for carbamyl phosphate in *E. coli* is in the formation of citrulline from ornithine in the pathway of L-arginine biosynthesis (Fig. 2.9). When arginine is in excess, ornithine is not produced because arginine exerts a specific inhibition on N-acetyl-L-glutamate synthetase [36], the first enzyme of the pathway leading to ornithine production (Fig. 2.9). Accordingly, when arginine is available for growth, carbamyl phosphate is used only for pyrimidine biosynthesis, and carbamyl phosphate synthetase and not ATCase becomes the first enzyme of the pyrimidine pathway. The specific inhibition by uridine nucleotides, particularly UMP *in vitro*, suggests that the primary control of pyrimidine nucleotide biosynthesis may be at this enzyme rather than ATCase, when arginine is in excess. However, when arginine is limiting, ornithine concentrations rise. Since ornithine is an allosteric activator of carbamyl phosphate synthetase and reverses UMP inhibition [33,35], the primary control of pyrimidine biosynthesis presumably reverts to ATCase under arginine limitation. This illustrates a very instructive principle of metabolic regulation. Although a pathway must be regulated by the enzyme catalysing the rate limiting step, this step may vary with the metabolic state of the cell.

The use of CTP as the feedback inhibitor of ATCase rather than UTP, initially seems curious, because if CTP was in excess but the uridine pool was limiting, ATCase would be inhibited and unable to provide the cell with UTP. Since CTP cannot be incorporated into RNA unless UTP is present in roughly equal amounts, growth would cease [28]. However, the detailed analysis of the inhibition of ATCase by CTP, shows that provided carbamyl phosphate is in excess, inhibition by CTP *in vitro* is incomplete, and never exceeds 85% [28]. Since a bacterium deficient in UMP and UTP should have high carbamyl phosphate levels (carbamyl phosphate synthetase not being inhibited) intermediates would continue to trickle through to UTP production, despite a great excess of CTP. There is currently no evidence for the existence of a second ATCase not inhibited by CTP and under the control of the uridine pool. This example clearly illustrates how a quantitative analysis of the action of an allosteric effector, can be crucial to an understanding of how a control mechanism operates *in vivo*.

2.9 The structure of ATCase: separate catalytic and regulatory subunits in a single enzyme

In a classical experiment, Gerhart and Schachman [37] were able to dissociate ATCase (sedimentation coefficient 11.7S) into two smaller units of 5.8S and 2.8S. The 5.8S component was catalytically active but not inhibited by CTP or activated by ATP, while the 2.8S component was catalytically inactive but possessed binding sites for CTP and ATP. Moreover, recombination of the 2.8S and 5.8S components restored the sensitivity of the latter to inhibition by CTP. It is now known that the 5.8S species ($M_r \sim 100\,000$) is a *trimer* composed of three identical catalytic (C) subunits, while the 2.8S species ($M_r \sim 34\,000$) is a *dimer* of two identical regulatory (R) subunits. The 11.7S ATCase enzyme ($M_r \sim 300\,000$) has the structure

R_6C_6 and its dissociation can be written by the following equation:

$$R_6C_6 \longrightarrow 3R_2 + 2C_3 \qquad [38]$$

The three dimensional structure of ATCase has been solved to 0.3 nm resolution by X-ray crystallography, and the arrangement of the subunits is shown in Fig. 2.12. The molecule shows D_3 molecular symmetry. The two C_3 *trimers* are in contact with each other, while the three R_2 *dimers* link catalytic chains in opposite C_3 *trimers*. Zinc atoms present on the boundaries between the R and C subunits play an essential role in the association of the enzyme [39].

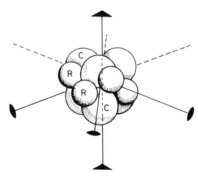

Fig. 2.12 View of the subunit structure of the aspartate transcarbamylase molecule perpendicular to the three-fold axis. Two-fold axes passing through the regulatory dimers are also shown. The catalytic and regulatory chains have been idealized as spheres for the sake of clarity. The triangles of three larger spheres represent the catalytic trimers, while the small spheres (shaded) represent the regulatory dimers.

Since the active and allosteric sites are actually located on different polypeptide chains, this proves that these sites are distinct, and strongly supports the allosteric concept. It is now clear that the activator ATP and inhibitor CTP bind at the same site on the R-subunit, and that this allosteric site is about 6 nm from the nearest catalytic centre [39]. Therefore ATP and CTP must affect catalysis by producing structural changes that are transmitted from the R to the C-subunits across the *subunit contacts* that link these two polypeptide chains together in the native enzyme. How this might be achieved is discussed in Chapter 6.

It should be emphasized, however, that the occurrence of distinct subunits for catalysis and regulation is not a general rule. Threonine deaminase and aspartokinase I—homoserine dehydrogenase I are each formed from four identical subunits, so that the active and allosteric sites are on the same polypeptide in these cases.

2.10 Summary

(a) Feedback inhibition by the end product on the first enzyme of a pathway, is a common mechanism by which the biosynthetic pathways in bacteria are regulated. Such a mechanism minimizes the requirement for carbon, nitrogen and energy and ensures that the whole pathway responds to changes in the ultimate product. It has been pointed out by Atkinson [40], that a further important reason against control at a later step is the marked increase in the concentration of intermediates before the regulated step that such a control would cause. The presence of thousands of poly-electrolytes and small metabolites in each cell poses serious problems of solubility, and the maintenance of low metabolite concentrations may be a compelling reason for citing control at the first committed step in a sequence.

(b) The two probes that have been especially powerful in elucidating these mechanisms are the analysis of the properties of isolated enzymes, and the behaviour of mutant bacteria with specific enzymic defects. These mutant studies have been crucial in establishing that regulatory mechanisms postulated from enzymatic and protein chemical analyses *in vitro*, actually function *in vivo*.

(c) The occurrence of branched pathways, and functional interdependence between pathways requires additional controls over the basic pattern of end product inhibition. These mechanisms of regulation and the structures of the enzymes into which the controls are inserted are quite diverse.

References

[1] Abelson, P. H. (1954), *J. Biol. Chem.*, **206**, 335–343.

[2] Umbarger, H. E. (1969), *Curr. Tops. Cell. Reg.*, **1**, 57–76.

[3] Umbarger, H. E. (1955), in *Amino Acid Metabolism*, 442–451, (eds McElroy and Glass), Johns Hopkins Press, Baltimore.

[4] Umbarger, H. E. (1956), *Science*, **123**, 848.

[5] Changeux, J. P. (1961), *Cold Spring Harb. Symp. Quant. Biol.*, **26**, 313–318.

[6] Monod, J., Changeux, J. P. and Jacob, F. (1963), *J. Mol. Biol.*, **6**, 306–329.

[7] Changeux, J. P. (1963), *Cold Spring Harb. Symp. Quant. Biol.*, **28**, 497–504.

[8] Freundlich, M. and Umbarger, H. E. (1963), *Cold Spring Harb. Symp. Quant. Biol.*, **28**, 505–511.

[9] Stadtman, E. R., Cohen, G. N., Le Bras, G. and Robichon-Szulmay-ster, H. (1961), *J. Biol. Chem.*, **236**, 2033–2038.

[10] Yugari, Y. and Gilvary, C. (1962), *Biochem. Biophys. Acta*, **62**, 612–614.

[11] Cohen, G. N. (1969), *Curr. Tops. Cell. Reg.*, **1**, 183–231.

[12] Patte, J. C., Le Bras, G. and Cohen, G. N. (1967), *Biochem. Biophys. Acta*, **136**, 245–257.

[13] Wormser, E. H. and Pardee, A. B. (1958), *Arch. Biochem. Biophys.*, **78**, 416–432.

[14] Rowbury, R. J. (1962), *Biochem. J.*, **82**, 24P.

[15] Brown, K. D. and Day, C. H. (1963), *Biochem. Biophys. Acta*, **77**, 170–172.

[16] Cotton, R. G. H. and Gibson, F. (1965), *Biochem. Biophys. Acta*, **100**, 76–88.

[17] Cohen, G. N. and Dautry-Varsat, A. (1980), in *Multifunctional Proteins*, (eds H. Bisswanger and E. Schmincke-Ott), John Wiley and Sons Inc., New York, ch. 3, pp. 49–121.

[18] Katinka, M., Cossart, P., Sibilli, L., Saint-Girons, I., Chalvignac, M. A., LeBras, G., Cohen, G. N. and Yaniv, M. (1980), *Proc. Nat. Acad. Sci* (USA), **77**, 5730—5733.

[19] Veron, M. and Cohen, G. N. (1972), *Eur. J. Biochem.*, **28**, 520—527.

[20] Truffa-Bachi, P. and Cohen, G. N. (1973), *Ann. Rev. Biochem.*, **42**, 113—134.

[21] Mally, M. I., Grayson, D. R. and Evans, D. R. (1981), *Proc. Nat. Acad. Sci* (USA), **78**, 6647—6651.

[22] Lumsden, J. and Coggins, J. R. (1978), *Biochem. J.*, **169**, 441—444.

[23] McCarthy, A. D. and Hardie, D. G. (1983), *Eur. J. Biochem.*, **130**, 185—193.

[24] Zocher, R., Keller, U. and Kleinkauf, H. (1982), *Biochemistry*, **21**, 43—48.

[25] Yourno, J., Kohno, T. and Roth, J. R. (1970), *Nature*, **228**, 820—824.

[26] Datta, P. and Gest, H. (1964), *Nature*, **203**, 1259—1261.

[27] Paulus, H. and Gray, E. (1964), *J. Biol. Chem.*, **239**, 4008—4009.

[28] Gerhart, J. C. (1970), *Curr. Tops. Cell. Reg.*, **2**, 276—325.

[29] Gerhart, J. C. and Pardee, A. B. (1962), *J. Biol. Chem.*, **237**, 891—896.

[30] Neuhard, J. (1968), *J. Bacteriol.*, **96**, 1519—1527.

[31] O'Donovan, G. A. and Neuhard, J. (1970), *Bact. Reviews*, **34**, 278—343.

[32] Long, C. W. and Pardee, A. B. (1967), *J. Biol. Chem.*, **242**, 4705—4721.

[33] Pierard, A. (1966), *Science*, **154**, 1572—1573.

[34] Anderson, P. M. and Meister, A. (1966), *Biochemistry*, **5**, 3164—3169.

[35] Ahmed, A. and Ingraham, J. L. (1969), *J. Biol. Chem.*, **244**, 4033—4038.

[36] Vyas, S., and Maas, W. K. (1963), *Arch. Biochem. Biophys.*, **100**, 542—546.

[37] Gerhart, J. C. and Schachman, H. K. (1965), *Biochemistry*, **4**, 1054—1062.

[38] Rosenbusch, J. P. and Weber, K. (1971), *J. Biol. Chem.*, **246**, 1644—1657.

[39] Kantrowitz, E. R., Pastra-Landis, S. C. and Lipscomb, W. N. (1980), *Trends in Biochemical Sciences*, **5**, 124—128 and 150—153.

[40] Atkinson, D. E. (1970), *Current Topics in Cellular Regulation*, **1**, 29—43.

3 The initiation of biological function by limited proteolysis

3.1 Zymogen activation in the small intestine

The activation of the pancreatic zymogens must rank as the first system in which the concept of enzyme regulation was clearly recognized. Even at the turn of the century, it was known that freshly secreted pancreatic juice only became able to digest proteins when it came into contact with a factor called enterokinase, in the small intestine. Further progress took place in two stages. In the 1930s, improved protein fractionation techniques enabled several of the enzymes and their inactive precursors to be *completely* resolved from one another, allowing an outline of the basic activation events to be formulated [1]. The introduction of more sophisticated separation techniques in the 1950s and the discovery of methods for solving the primary and tertiary structures of proteins then allowed the detailed molecular events which accompany activation to be analysed.

3.1.1 The pancreatic proteinases and their specificities

Activated pancreatic juice contains at least seven distinct proteinases, (Table 3.1). There are three types of endopeptidase, trypsins (T) chymotrypsins (ChT), and elastases (E), and two types of exopeptidase, carboxypeptidase A (CpA) and carboxypeptidase B (CpB). Since the endopeptidases have specificities that complement one another, their combined action will lead to extensive fragmentation of dietary proteins which have already been partially hydrolysed by proteinases in the stomach. The exopeptidases, CpA and CpB, catalyse the sequential release of amino acids from the C-termini of peptides: CpB removing C-terminal arginine and lysine from peptides generated by tryptic attack, and CpA, C-terminal amino acids from peptides generated by digestion with chymotrypsin and elastase, or

Table 3.1 Specificities of the pancreatic proteinases.

Proteinase	Type	Specificity	Peptide bonds cleaved
trypsins I and II	endopeptidase	narrow	*lys*-X, *arg*-X
chymotrypsins A, B, C	endopeptidase	broad	*tyr*-X, *phe*-X, *trp*-X, *leu*-X, *his*-X
elastases I and II	endopeptidase	broad	*val*-X, *ala*-X, *gly*-X, *ser*-X
carboxypeptidase A	exopeptidase	very broad	X-*tyr*, X-*phe*, X-*trp*, X-*leu*, X-*ile*, X-*met*, X-*his*, X-*ala*, X-*val*, X-*thr*
carboxypeptidase B	exopeptidase	narrow	X-*lys*, X-*arg*

tryptic peptides already attacked by CpB. Further digestion is carried out by a variety of peptidases secreted into the small intestine, from intestinal cells. A proline dipeptidase which specifically hydrolyses x-pro dipeptides is particularly important [2] since pro-x and x-pro bonds are resistant to cleavage by pancreatic proteinases. Another is the classical 'leucine aminopeptidase', which catalyses the sequential release of amino acids from the N-termini of peptides [3].

3.1.2 Secretion and activation of the pancreatic zymogens

All seven proteinases are stored as inactive precursors termed *zymogens*; trypsinogen (Tg), chymotrypsinogens (ChTg) A, B and C, proelastase (ProE), and procarboxypeptidases A and B (ProCpA and ProCpB). They are located in the *acinar* cells of the *exocrine* pancreas within spherical *zymogen granules*, separated from the rest of the cell constituents by a membrane. Secretion into the duodenum takes place by *exocytosis*, a process in which the zymogen granule membrane fuses with the outer membrane of the *acinar* cell, and which is essentially the reverse of phagocytosis. It is stimulated by excitation of the vagal nerve to the pancreas and by the peptide hormones, *cholecystokinin* and *secretin*. The latter are synthesized in intestinal cells in the upper region of the small intestine, and are secreted into the blood circulation, from where they migrate to the pancreas. Cholecystokinin stimulates the secretion of a small quantity of pancreatic juice rich in enzyme precursors, while secretin stimulates a copious flow of juice, rich in HCO_3^- ions, but low in zymogens. The HCO_3^- ions are essential to buffer the acid contents of the stomach, since pancreatic proteinases have rather sharp pH-activity profiles with optima near pH 8.0. Among the most potent stimulators of cholecystokinin secretion are certain L-amino acids and long chain fatty acids, while acid itself is the most effective stimulator of secretin release [4–6].

Enteropeptidase (Ep), originally called enterokinase, is a proteolytic enzyme synthesized in the brush border cells of the small intestine. Its presence in the intestinal tract starts the process of zymogen activation (Fig. 3.1) by initiating the conversion of trypsinogen to trypsin [1]. Its

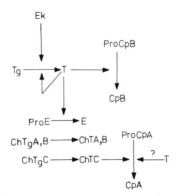

Fig. 3.1 Probable sequence of events during activation of the pancreatic zymogens.

crucial role *in vivo* is demonstrated by the severe intestinal disorders which occur in human patients who lack enteropeptidase activity [7]. Once initiated, the conversion of trypsinogen to trypsin becomes autocatalytic, since trypsin can catalyse its own formation from trypsinogen (Fig. 3.2). Trypsin also catalyses the activation of proelastase [8], the chymotrypsinogens [9–11] and ProCpB [12]. ProCpA exists as a complex with chymostrpsinogene, and probably with a further zymogen, called zymogen E. Its mechanism of activation is likely to involve the formation of chymotrypsin C by trypsin, followed by the activation of ProCpA by chymotrypsin C [13].

Each zymogen is activated by the cleavage of just a single peptide bond (Table 3.2). Enteropeptidase cleaves the susceptible *lys* 6-*val* 7 bond in trypsinogen 2000-fold more rapidly than equivalent amounts of trypsin, presumable because it recognizes the unusual $(asp)_4$-*lys* sequence [14]. Nevertheless, the autocatalytic activation of trypsinogen by trypsin is still likely to be physiologically significant, since there are only trace amounts of enteropeptidase the small intestine relative to trypsin. Enteropeptidase does not activate any other zymogen.

Although full chymotryptic activity is generated by the tryptic cleavage of a single *arg* 15-*ile* 16 bond, several further peptide bonds (*leu* 13-*ser* 14, *tyr* 146-*thr* 147 and *asn* 148-*ala* 149) are then cleaved autocatalytically by chymotrypsin, but without any further change in activity. Two dipeptides *ser-arg* and *thr-asn* are released, but the three polypeptide chains of chymotrypsin A do not dissociate as they are cross-linked by disulphide bridges [9].

The structure of the activation peptides of proelastases I and II from rat pancreas have recently been elucidated by sequencing the cDNA prepared to the messenger RNA of these enzymes. The activation peptide of proelastase-II is similar to that of trypsinogen and chymotrypsinogen (Table 3.2), whereas that of proelastase-I is quite different [15].

The activation of ProCpA by chymotrypsin C involves the cleavage of a single peptide bond and the loss of 94 amino acids from the N-terminus

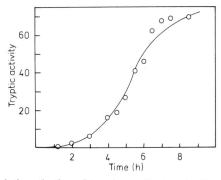

Fig. 3.2 Autocatalytic activation of trypsinogen by trypsin. Open circles are experimental points, and the full line a theoretical curve for a simple autocatalytic reaction [1].

Table 3.2 Activation of some pancreatic zymogens. Peptide bonds cleaved during the activation process are denoted by arrows [8,11,15,17]

Zymogen	Activation peptide
Bovine Tg	val-asp-asp-asp-asp-lys $\downarrow$ ile-val-
Porcine Tg	phe-pro-thr-asp-asp-asp-asp-lys $\downarrow$ ile-val-
Bovine ChTg-A	cys-gly-val-pro-ala-ile-gln-pro-val-leu-ser-gly-leu-ser-arg $\downarrow$ ile-val-
Bovine ChTg-B	cys-gly-val-pro-ala-ile-gln-pro-val-leu-ser-gly-leu-ala-arg $\downarrow$ ile-val-
Bovine ChTg-C	cys-gly-ala-pro-ile-phe-gln-pro-asn-ser————ala-arg $\downarrow$ ile-val-
Rat Pro-E II	cys-gly-tyr-pro-thr-tyr-gln-val-gln-his —asp-val-ser-arg $\downarrow$ val-val-

[16]. Although CpB and CpA are almost identical in size ($M_r \sim 35\,000$), ProCpB (M_r 57 000) is much bigger than ProCpA, implying that activation involves the removal of a much larger fragment. Tryptic cleavage of a single *arg-thr* bond generates 60–70% of maximal CpB activity without a significant decrease in size. Maximal activity which is generated more slowly, is accompanied by a reduction to an M_r of 35 000, and may involve further peptide bond cleavage [12].

The synthesis of proteinases as zymogens is presumably essential to prevent them from degrading other pancreatic proteins before they are packaged into zymogen granules, and from digesting one another within the zymogen granules before they are secreted. The use of specific secretory granules may facilitate the control of secretion, and also prevent the zymogens from being activated by other proteinases present within the acinar cells, but outside the granules. There is a further protein secreted in pancreatic juice, which is a specific inhibitor of trypsin [1,18]. Trypsin inhibitor comprises only 2% of the protein content of pancreatic juice, very much less than the concentration of trypsinogen [19]. Its function is presumably to prevent autocatalytic activation of trypsinogen by any traces of trypsin that might be present in granules, ensuring that activation of zymogens only takes place when pancreatic juice meets enteropeptidase.

The activation of the pancreatic zymogens is an example of the use of enzyme regulation as a means of amplifying the effect of an initial stimulus. Each molecule of enteropeptidase converts many molecules of trypsinogen to trypsin; each molecule of trypsin generates more molecules of trypsin; each molecule of trypsin activates other zymogens, and each molecule of activated proteinase hydrolyses many peptide bonds. Further examples of this phenomenon will be found later in this chapter and in Chapters 4 and 5. The location of each activation site near the amino-terminus of each protein, and the need to cleave just a single bond to generate activity are good examples of biological economy. The native configuration of zymogens clearly allows trypsin to cleave one particular *arg*-X or *lys*-X bond preferentially, but prevents it from attacking the many other potentially susceptible linkages, which would be accessible, if the zymogen was in the denatured state.

It is not commonly realized that the human pancreas secretes about 10 grams of hydrolytic enzymes each day, which is a significant proportion of the average daily intake of protein of some diets [20]. Consequently, it is an important facet of the control process, that the proteinases should be sufficiently susceptible to proteolysis that they eventually degrade each other to their constituent amino acids which are then reabsorbed.

3.1.3 The conversion of chymotrypsinogen to chymotrypsin

The tertiary structures of both chymotrypsin and chymotrypsinogen have been solved to 0.25 nm resolution by X-ray crystallography [21,22]. The residue essential for catalysis in chymotrypsin is *ser* 195. This is hydrogen bonded to *his* 57 which is in turn hydrogen bonded to *asp* 102 (Fig. 3.3). The series of hydrogen bonds has been described as a charge relay system in which the negative charge is transferred from the buried residue *asp* 102, through a hydrophobic environment to the serine oxygen. This increases the electronegativity of the oxygen atom, making the normally inert serine hydroxyl a powerful nucleophile capable of initiating a nucleophilic attack on the carbonyl carbon of the peptide bond to be hydrolysed. The importance of *ser* 195 had of course been realized some years prior to the crystallographic analysis, since it reacts rapidly with di-isopropyl-fluoro phosphate (DFP) to form a covalent derivative that is totally inactive. Trypsin, elastase and even enteropeptidase are also inactivated by DFP. The significance of this observation is discussed below.

The overall folding of chymotrypsinogen and chymotrypsin is very similar and a comparison of the two structures suggests that only a few significant movements of the chain take place upon activation. The most surprising observation is that the positions of *ser* 195, *his* 57 and *asp* 102 are virtually unchanged in the zymogen and activated enzyme [22]. This is consistent with observations that *ser* 195 has a special reactivity in the zymogen, although it reacts considerably more slowly with DFP or with cyanate than the activated enzyme [23].

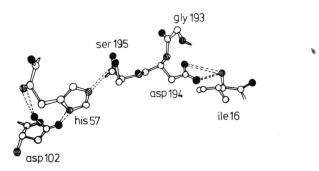

Fig. 3.3 The conformation of some important amino acids at the active centre of chymotrypsin. The viewpoint is outside the surface of the molecule working towards the interior. o-carbon, o-nitrogen, •-oxygen atom [10].

From the standpoint of activation, the following events appear to be crucial. The free α-amino group of *ile* 16 generated by tryptic cleavage (Fig. 3.2) tucks inside the molecule where it forms an ion-pair with *asp* 194 (Fig. 3.3). As a result the side chains of *ile* 16 and *val* 17 also become buried, and *asp* 194 no longer hydrogen bonds to *his* 40 as it does in the zymogen. *Gly* 193 forms a new hydrogen bond with *his* 40. *Arg* 145 moves from the surface to the interior, where it may interact with the carboxylate side chain of *asp* 194. *Met* 192 moves from the interior, to the surface site occupied by *arg* 145 in the zymogen. These alterations form the four new specificity determinants, which are missing in the zymogen, out of the nine by which chymotrypsin recognises its polypeptide substrates. It has been estimated that this could account for most of the 10^4–10^6 greater activity of chymotrypsin compared to the zymogen [24]. A slight rearrangement of *gly*-193, which is important in stabilising the transition state, may also promote catalysis by the enzyme [24].

The conversion of chymotrypsinogen to chymotrypsin is a good illustration of the sort of structural changes that might take place during an allosteric transition (Chapter 2). The cleavage of a peptide bond at a surface site well removed from the active centre (analogous to the binding of an allosteric effector) initiates structural changes which lead to the formation of the substrate binding pocket. There are no gross changes in the overall structure, and relatively small shifts in the position of just a few of the 230 amino acids are sufficient to accomplish the transition (see Chapter 6).

Although the activation peptides of trypsinogen and chymotrypsinogen do not closely resemble one another (Table 3.2), trypsin, elastase and chymotrypsins A and B each comprise about 230 amino acids and are all specifically inhibited by DFP. The *overall* sequence identity is 78% between chymotrypsins A and B, and 40–55% between any other pair of enzymes [8]. The sequences in the vicinity of the critical amino acids are very similar (Table 3.3). The tertiary structures of trypsin [25] and elastase [8] show the same overall folding as chymotrypsin A, and the charge relay system (Fig. 3.3) is identical, indicating a common catalytic mechanism for each enzyme. The different specificities of each enzyme are simply explained by a few particular replacements in the substrate binding pockets. The four enzymes were clearly derived from a common ancestral gene and form part of a more extensive family of structurally related serine proteinases (Table 3.3, Sections 3.2, 3.3). These analyses also suggest that it is relatively much simpler to evolve a different substrate specificity than a new mechanism of catalysis. This may well turn out to be a general rule of enzyme evolution.

Crystallographic evidence has shown that the activation of trypsinogen is similar to that of chymotrypsinogen [26]. Circular dichroision studies of acyl intermediates formed during catalysis by trypsin and trypsinogen have demonstrated that the substrate is bound differently in enzyme and zymogen. Interestingly, addition of the N-terminal dipeptide *ile–val* of trypsin (Table 3.3) to trypsinogen converts the latter protein to a trypsin like conformation, while conversely, blocking the N-terminal *ile*-16 transforms trypsin to a trypsinogen-like conformation [26,27].

Table 3.3 Sequence homologies between some mammalian serine proteinases. Residue numbers refer to chymotrypsinogen. A—amino terminal sequences; B,C,—sequences surrounding residues of the charge relay system.

Enzyme	A	B	C	Ref
	16	57	195	
Bovine Chymotrypsins	ile-val-asn-gly	trp-val-val-thr-ala-ala-his-cys-gly	cys-met-gly-asp-ser-gly-gly-pro-leu	[8]
Bovine Trypsin	ile-val-gly-gly	trp-val-val-ser-ala-ala-his-cys-try	cys-gln-gly-asp-ser-gly-gly-pro-val	[8]
Porcine Elastase	val-val-gly-gly	trp-val-met-thr-ala-ala-his-cys-val	cys-gln-gly-asp-ser-gly-gly-pro-leu	[8]
Bovine Thrombin	ile-val-glu-gly	trp-val-leu-thr-ala-ala-his-cys-leu	cys-glu-gly-asp-ser-gly-gly-pro-phe	[39]
Bovine Factor XIIa	val-val-gly-gly	val-leu-thr-ala-ala-his-cys-leu*	cys-gln-gly-asp-ser-gly-gly-pro-leu	[32]
Bovine Factor IXa	val-val-gly-gly	trp-val-val-thr-ala-ala-his-cys-ile	cys-gln-gly-asp-ser-gly-gly-pro-his	[41]
Bovine Factor Xa	ile-val-gly-gly	tyr-val-leu-thr-ala-ala-his-cys-leu	cys-gln-gly-asp-ser-gly-gly-pro-his	[40]
Human C1s̄	ile-ile-gly-gly	trp-val-leu-thr-ala-ala-his-val	cys-lys-gly-asp-ser-gly-gly-ala-phe	[54,55]
Human C1r̄	ile-ile-gly-gly	trp-ile-leu-thr-ala-ala-his-thr	cys-gln-gly-asp-ser-gly-gly-val-phe	[55]

* K. Fujikawa and E. W. Davie, personal communication.

31

The two isoenzymes carboxypeptidase A and B show a 51% identity in amino acid sequence [28]. The tertiary structure of carboxypeptidase A [17] but not carboxypeptidase B, has been solved to 0.2 nm resolution, but the conformation of the zymogens and hence the structural changes that accompany activation, are unknown.

3.1.4 Secretion and synthesis

Although the structures of the pancreatic proteinases and their precursors are now understood in great detail, in the last analysis, the real control is exerted by enteropeptidase, secretin and cholecystokinin. It was the discovery of secretin in 1902 [29] which led to the introduction of the term hormone to describe a molecule which influenced the metabolism of a cell distinct from that in which it was synthesized. However, the mechanisms by which these hormones promote exocytosis, or even the molecular events which take place during exocytosis are unclear.

A considerable amount of information about the synthesis of pancreatic zymogens has been obtained from cell free translation systems. Like most other secreted proteins the pancreatic zymogens are synthesized as *preproenzymes*, with an N-terminal extension of some 20 residues that is absent in the mature zymogen (e.g. [15]). This so called 'signal sequence' which contains a high proportion of non-polar residues, serves to direct the nascent protein across the endoplasmic reticulum (ER) membrane. The signal sequence is removed by a signal peptidase on the luminal side of the ER, and the mature zymogen is then transferred to the *Golgi apparatus* where packaging into the granules takes place. Signal peptidases clearly constitute another important general mechanism by which biological functions can be induced by limited proteolysis [30,31].

3.2 The molecular basis of blood coagulation

When the blood vessels in a tissue become damaged, three factors act to minimize the loss of blood; vasoconstriction of blood vessels, platelet aggregation and blood clotting at the site of injury. Although the control of blood clotting is not completely understood, progress has been particularly marked in recent years. This stems from the fact that all of the factors, some of which are present in only trace amounts, have only recently been obtained in highly purified, undergraded forms, that are completely free from one another. This has enabled their interactions with one another, and activation mechanisms to be studied unambiguously [32,33]. This section provides a brief synopsis of current views about this classical system of enzyme regulation by limited proteolysis.

Two distinct mechanisms of blood coagulation have been observed *in vitro*, termed the *intrinsic* and *extrinsic* pathways (Figs. 3.4, 3.5). The intrinsic system uses only factors found in blood plasma, while the *extrinsic* system uses factors present in both plasma and tissue extracts. Although they can be separated *in vitro*, both mechanisms are important *in vivo*, as judged by bleeding disorders in human patients who lack one of these factors [34]. These genetic mutants have been invaluable in identifying and ordering the steps in these regulatory pathways. The two pathways

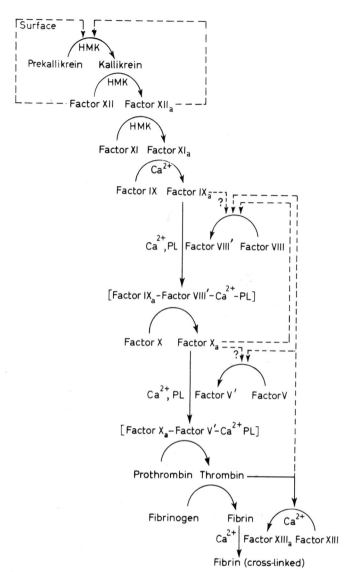

Fig. 3.4 Tentative mechanism for the initiation of blood clotting in mammalian plasma in the intrinsic system [32]. PL-phospholipid; a –activated form of clotting factor; HMK, high molecular weight kininogen.

probably function together, but to different extents *in vivo* depending on the particular conditions that prevail.

In the intrinsic pathway, a number of clotting factors act upon one another in a stepwise manner, eventually leading to the conversion of the protein fibrinogen to fibrin. Fibrin undergoes both end to end and lateral aggregation to form an insoluble fibrin clot, which is stabilized by factor

33

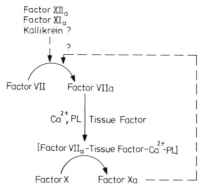

Fig. 3.5 Possible mechanism for initiation of blood clotting through the extrinsic system. Abbreviations as in Fig. 3.4 [32].

XIIIa. This is a transglutaminase enzyme, which catalyses the formation of $\epsilon(\gamma$-glutamyl) lysine cross-linkages between glutamine side chains on one fibrin molecule and lysine side chains on another.

The intrinsic pathway is the classical example of the amplification phenomenon (Section 3.1), and can partly explain the extremely sudden formation of the fibrin clot after a brief lag period. However, two additional factors which also contribute to the speed of the response, are the *feedback activation* of factor VIII and factor V by the first traces of factor Xa and thrombin that are formed (Fig. 3.4), and the reaction of factors at a phospholipid surface, which increases their effective concentration. The extrinsic pathway bypasses several of the clotting factors and meets the intrinsic pathway at the level of factor X. The two pathways are identical after this point. Factor X is therefore activated by two distinct clotting factors *in vivo*.

Factors XII, XI, IX, X, VII and prothrombin are inactive precursors of proteolytic enzymes, which catalyse the activation of the next factor in the sequence, when provided with appropriate cofactors. Factor XIII is the precursor of a transglutaminase enzyme, but factors VIII, V and fibrinogen and perhaps tissue factor may not be precursors of enzymes. Nevertheless, they are converted to active or more active proteins by limited proteolysis.

3.2.1 The intrinsic and extrinsic pathways
The intrinsic pathway is initiated *in vitro* by the contact of factor XII with a surface. This is normally the glass wall of a reaction vessel *in vitro*, but *in vivo* may well be fibrous proteins, such as collagen or elastin, which surround the damaged blood vessel. Surface contact promotes a conversion to the activated form, factor XIIa, but this occurs only very slowly *in vitro*. Factor XIIa in the presence of high molecular weight kininogen has been shown to convert prekallikrein to kallikrein *in vitro* by limited proteolysis. Kallikrein is a proteinase in blood plasma, one of whose functions is thought to be the generation of pharmacologically active peptides, such as bradykinin, from high molecular weight kininogen. Bradykinin has several effects; for example it stimulates relaxation of the walls of blood vessels

and thus promotes vasodilation [35]. Kallikrein has also been shown to convert factor XII to XIIa *in vitro* suggesting a possible mechanism for amplifying the activation of factor XII, once the first traces of factor XIIa have been formed (Fig. 3.4). However, these early stages in the blood clotting sequence are not fully understood. People who lack factor XII are usually discovered by chance and have no serious bleeding disorders [34], and indeed the original person in whom the factor XII defect was identified (Hageman) died from a pulmonary thrombosis [36]!

Factor XIa converts factor IX to IXa. Factor IXa cannot by itself activate factor X, but must first form a complex with factor VIII. Current evidence suggests that factor VIII supports the factor IXa catalysed activation of factor X very poorly (or perhaps not at all) unless it is converted to factor VIII'. This can be carried out *in vitro* by thrombin and factor Xa, but where these activated proteinases would come from is unclear.

Human patients lacking a functional factor VIII have the disease known as haemophilia. This is the most common blood clotting disorder, occurring in 1 in every 10 000 births [34]. Since the gene for haemophilia is located on the X-chromosome, females will generally only be carriers of the disease, and the vast majority of haemophiliacs are males. The most distinguished carrier was Queen Victoria. Her fifth son Leopold suffered from haemophilia and died at the age of 31 from a brain haemorrhage after a minor blow on the head. Two of Queen Victoria's daughters were also carriers of haemophilia, and passed the disease to other royal families in Europe. This was particularly important in the case of the Czarist succession in Russia, since the only son of the marriage between Queen Victoria's daughter Alice and Czar Nicholas II, was a haemophiliac who died at the age of 14 [37].

The conversion of prothrombin to thrombin catalysed by factor Xa and factor V is closely analogous to the activation of factor X by the factor IXa–factor VIII' complex, with factor V playing the same role as factor VIII'. Like factor VIII, factor V activity is greatly increased by conversion to factor V'. This reaction is catalysed *in vitro* by thrombin and perhaps factor Xa. The purification and characterization of factors V and VIII remains one of the major challenges in the study of blood coagulation.

The mechanism of conversion of factor X to Xa by the extrinsic system is still not fully understood. It has become clear that factor VII can be converted to a proteolytic enzyme (factor VIIa), but the conversion of factor X to Xa still requires tissue factor [38]. Tissue factor may possibly play a role analogous to that of Factor V' and factor VIII'. Factor VII can be converted to factor VIIa *in vitro* very effectively by factor Xa. However, the factor which might initiate the activation of factor VII *in vivo* is an open question, although kallikrein and factor XIIa have been suggested.

The conversion of factor X to Xa by factor IXa–factor VIII' or factor VIIa–tissue factor and the conversion of prothrombin to thrombin by factor Xa – factor V' have an absolute requirement for Ca^{2+} and phospholipid while the conversion of factor IX to IXa by factor XIa requires Ca^{2+}. A beautiful molecular explanation for these requirements has been discovered and is described in Section 3.2.2.

3.2.2 Molecular events in the activation of the blood clotting factors

The proteolytic enzymes, factor XIIa, factor XIa, factor IXa, factor VIIa, factor Xa and thrombin are all inhibited by DFP, which reacts with a special serine residue. The amino acid sequences surrounding this serine and the other active site regions of factor IXa, Xa, XIIa, and thrombin are almost identical with one another and also with the pancreatic endopeptidases. Furthermore, the N-terminal sequences generated upon activation of the zymogens are also homologous with the pancreatic proteinases Table 3.3. These results, coupled with the complete amino acid sequence of thrombin [39], factor X [40] and factor IX [41] show that the activated forms of these blood coagulation factors (which are synthesized in the liver) and the pancreatic proteinases have all evolved from the same ancestral protein, and act by the same catalytic mechanism. They also suggest that once the susceptible bond in each zymogen has been cleaved, the structural transitions which lead to the generation of activity may also be very similar.

The conversion of factor X to Xa catalysed by factor IXa or factor VIIa involves the cleavage of the same *arg-ile* bond [42]. The conversion of factor IX to IXa by factor XIa involves an initial cleavage of an *arg-ala* bond without activation followed by the cleavage of an *arg-val* bond with activation. The conversion of prothrombin to thrombin requires the initial cleavage of an *arg-thr* bond without activation, followed by an *arg-ile* bond with activation. Thrombin cleaves four *arg-gly* bonds during the fibrinogen to fibrin conversion and a single *arg-gly* bond in the conversion of factor XIII to XIIIa. Each activation is therefore achieved by the rupture of very few peptide bonds. The specificities of the enzymes are all very similar to trypsin, and indeed trypsin can mimic the action of factor IXa and factor X_a. The blood clotting factors are, however, extremely specific proteinases, which can cleave just one or two bonds in one or two proteins. The reason may lie in the recognition of a longer sequence than merely an *arg-X* bond, as in the case of enteropeptidase. It is therefore of interest that the two factor Xa sensitive bonds in prothrombin are preceded by the identical *ile-glu-gly-arg* sequence [43].

Prothrombin is one of four coagulation factors (factors IX, VII and X) which require vitamin K for their biosynthesis [44]. Vitamin K deficient animals synthesize an abnormal prothrombin, which cannot bind Ca^{2+} and cannot be activated *in vivo*, although it can be activated by trypsin *in vitro*, showing that the active centre is not impaired. It has now been discovered that normal prothrombin contains an amino acid, γ-carboxyglutamic acid (*Gla*), never previously found in proteins, at positions 7 and 8, 15 and 17, 20 and 21, 26 and 27 and 30 and 33 from the N-terminus [45]. These residues are glutamic acid in vitamin K deficient animals. This explains the abnormal Ca^{2+} binding, since Ca^{2+} binds to two adjacent *Gla* residues in a manner analogous to the binding of calcium to EDTA. Phospholipid binds to prothrombin through the Ca^{2+} which are bound to *Gla* and this enables the factor Xa-factor V' complex to bind to prothrombin and initiate activation. Factor X and Factor IX also have *Gla* residues in analogous positions and activation of these zymogens takes place by a mechanism perfectly analogous to that of prothrombin [42]. These

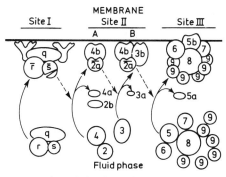

Fig. 3.6 Pictorial representation of the classical complement system; q, r and s refer to the subunits of the C1 complex; a, b represent the N− and C−terminal fragments of modified factors respectively. Proteinases are denoted by horizontal bars [50].

results have given an exciting clue to the function of vitamin K, which is essential for the activity of an enzyme system which converts glutamic acid to γ-carboxyglutamic acid [44].

Three further Vitamin K dependent proteins that contain *Gla* residues have been identified in blood, termed protein C [46], protein S [47] and protein Z [48]. Although the functions of these proteins are unknown, protein C has been shown to be converted to an active proteinase by thrombin. This proteinase has *anticoagulant* activity since it can degrade factor V and factor VII [49].

3.3 Activation of the complement system [50–53]

When a mammal is infected with a foreign organism such as an invading bacterium, its first response is to produce a specific antibody, which forms an immune complex with the invading cell. The next response involves complement, a series of blood plasma proteins which first recognize immune complexes and then cause structural damage to the invading cell. The third phase involves phagocytosis of the damaged cells by leukocytes, monocytes and macrophages in blood plasma. In the classical complement pathway (Fig. 3.6) the complement components C1r, C1s and C2 are precursors of proteinases, while C3, C4 and C5 are converted to functional derivatives by limited proteolysis.

The sequence is initiated when the C1 complex recognizes an immune complex between an invading cell and an IgG or IgM immunoglobulin. The recognition is effected by C1q which binds specifically to the antibody. It is presumed that this interaction induces a change in the conformation of C1q, which in turn induces a conformational change in C1r, converting it to a proteinase C1r̄. C1r̄ then cleaves a single peptide bond in C1s, converting it to the proteinase C1s̄. C1r̄ and C1s̄ are inhibited by DFP and the amino acid sequences around the modified serines indicate that they are members of the family of serine proteinases (Table 3.3).

C1s̄ cleaves peptide bonds in C4 and C2. The C-terminal fragment of C4, C4b, can then bind at another site on the membrane of the invading cell. The C-terminal fragment of C2, C2a becomes a proteinase only when

37

it is complexed with C4b. The C2$\bar{\text{a}}$–C4b complex then cleaves C3 [56]. A C2$\bar{\text{a}}$–C4b–C3b complex is required to cleave C5. The C-terminal fragment of C5, C5b, has the ability to trigger the formation of a complex between all the rest of the complement components at a third site on the membrane, without any further proteolysis. The insertion of the C5b–9 complex into the membrane in some way breaks the orderly structure of the membrane causing it to become leaky.

The C2$\bar{\text{a}}$–C4b and C2$\bar{\text{a}}$–C4b–C3b complexes are examples of highly specialized proteinases. The only known substrate for the former is C3 and for the latter C5. C2$\bar{\text{a}}$ is therefore a remarkable proteinase whose specificity is determined by the presence of two regulatory subunits C4b and C3b.

The binding of C3b and C4b to the cell membrane also seems to facilitate phagocytosis of complement coated particles by leucocytes, monocytes and macrophages. In addition, the fragments C3a and C5a which are not used directly in the complement system (Fig. 3.6) have several biological actions of their own; for example, they cause smooth muscle contraction and C5a stimulates the chemotactic migration of leucocytes.

A second complement system exists in blood plasma, which enters the classical pathway at the level of C3. This alternative pathway is activated by immune complexes which involve IgG and IgA immunoglobulins, and also by some non-immunoglobulin substances such as yeast or bacterial cell walls.

The C3 cleaving enzyme of the alternative pathway is a complex of a proteinase termed *factor B* and C3b. This unusual situation where the product of the reaction (C3b) is also part of the enzyme that catalyses the reaction means that the alternative pathway is self-amplifying. Factor B is activated on cleavage by the serine proteinase, factor D. Factor D may be unique among plasma serine proteinases because it has no known zymogen, and circulates as the active enzyme. This is possible, because factor D can only cleave factor B when it is complexed with C3b. Activation of the alternative pathway of complement therefore requires that C3 be cleaved by the classical pathway, or by some as yet undiscovered mechanism.

Factor B of the alternative pathway and C2 of the classical pathway are members of a new family of serine proteinases. They are much larger than the proteinases of the trypsin family [57] and lack the characteristic N-terminal sequences (Table 3.3), but have similar sequences around the histidine and serine residues of the charge relay system.

Recently, a new serine proteinase has been characterized called C3b/C4b inactivator or factor I, which destroys by limited proteolysis the biological activities of C3b and C4b [58]. The specificity of factor I, is determined by the presence of other regulatory proteins. In the presence of Factor H it cleaves C3b relatively specifically. However, when complexed with a different plasma protein, termed C4b binding protein, factor I cleaves C4b preferentially [59].

3.4 Summary

'Limited proteolysis is a highly specific irreversible process, which can serve to initiate physiological function by converting a precursor protein into a biologically active form' [11]. It can also be a mechanism for terminating biological activity. Limited proteolysis is used to regulate a wide range of processes in eukaryotic cells, including not only digestion, blood coagulation and defence, but blood pressure [60] the processing of a number of peptide hormones [61] fertilization [62] and the synthesis of connective tissue [63].

Most enzymes that are synthesized as precursors are proteinases. However, three which are not, are pancreatic phospholipase A [64], factor XIIIa, and chitin synthetase [65], which forms the N-acetyl glucosamine polymer in budding yeast.

Genetic variants of mammalian enzymes, which are often responsible for specific hereditary diseases in human patients, have been invaluable in establishing pathways of limited proteolysis *in vivo* as was the case for the bacterial enzymes discussed in Chapter 2.

In view of the major advances taking place in the fields of blood coagulation and complement, the rapidly growing number of proteins that are being found to be synthesized as inactive precursors, and the current interest in the role of proteinases in the regulation of protein and enzyme turnover in eukaryotic cells, the control of enzyme activity by limited proteolysis is likely to remain an important area of research for many years.

References

[1] Northrop, J. H., Kunitz, M. and Herriott, R. M. (1948), in *Crystalline Enzymes*, Columbia University Press, New York.

[2] Sjöström, H. and Noren, O. (1974), *Biochem. Biophys. Acta*, **359**, 177–185.

[3] Delange, R. J. and Smith, E. L. (1971), in *The Enzymes* III, (3rd Edition), (ed. P. D. Boyer), Academic Press, London, pp. 81–118.

[4] Wormsley, K. G. (1979), in *Scientific Basis of Gastroenterology*, (eds H. L. Duthie and K. G. Wormsley), Churchill-Livingstone, Edinburgh, pp. 163–248.

[5] Johnson, L. R. (1974), in *Gastrointestinal Physiology – MTP International Review of Science*, (eds E. D. Jacobson and L. L. Shanbour), Butterworths, London and University Park Press, Baltimore, pp. 1–43.

[6] Preshaw, R. M. (1974), in *Gastrointestinal Physiology – MTP International Review of Science*, (eds E. D. Jacobson and L. L. Shanbour), Butterworths, London and University Park Press, Baltimore, pp. 265–291.

[7] Hadorn, B., Tarlow, M. J., Lloyd, J. K. and Wolff, O. H. (1969), *The Lancet*, pp. 812–813.

[8] Hartley, B. S. and Shotten, D. M. (1971), in *The Enzymes* III, (3rd Edition), (ed P. D. Boyer), Academic Press, London, pp. 323–373.

[9] Kraut, J. (1971), in *The Enzymes* III, (3rd Edition), (ed. P. D. Boyer), Academic Press, London, pp. 165–183.

[10] Blow, D. M. (1971), in *The Enzymes* III, (3rd Edition), (ed. P. D. Boyer), Academic Press, London, pp. 185–212.

[11] Neurath, H., Walsh, K. A. and Gertler, A. (1974), in *Metabolic Interconversions of Enzymes* (1973), (eds. E. H. Fischer, E. G.

Krebs, H. Neurath and E. R. Stadtman), Springer-Verlag, Heidelberg, pp. 301–312.

[12] Wintersberger, E., Cox, D. J. and Neurath, H. (1962), *Biochemistry*, **1**, 1069–1078.

[13] Kobayashi, Y., Kobayashi, R. and Hirs, C. H. W. (1981), *J. Biol. Chem.*, **256**, 2466–2470.

[14] Maroux, S., Baratti, J. and Desnuelle, P. (1971), *J. Biol. Chem.*, **246**, 5031–5039.

[15] Macdonald, R. J., Swift, G. H., Quinto, C., Swain, W., Pictet, R. L., Nikovits, W. and Rutter, W. J. (1981), *Biochemistry*, **21**, 1453–1463.

[16] Quinto, C., Quiroga, M., Swain, W. F., Nikovits, W. C., Standring, D. N., Pictet, R. L., Valenzuela, P. and Rutter, W. J. (1982), *Proc. Nat. Acad. Sci* (USA), **79**, 31–35.

[17] Hartsuck, J. and Lipscomb, W. N. (1971), in *The Enzymes* III, (3rd Edition), (ed. P. D. Boyer), Academic Press, London, pp. 1–56.

[18] Ruhlmann, A., Kukla, D., Schwager, P., Bartels, K. and Huber, R. (1973), *J. Mol. Biol.*, **77**, 417–436.

[19] Keller, P. J. and Allan, B. J. (1967), *J. Biol. Chem.*, **242**, 281–287.

[20] Banks, P., Bartley, W. and Burt, L. M. (1976), *The Biochemistry of the Tissues*, 2nd Edition, John Wiley and Son, London.

[21] Sigler, P. B., Blow, D. M., Matthews, B. W. and Henderson, R. (1968), *J. Mol. Biol.*, **35**, 143–164.

[22] Freer, S. T., Kraut, J., Robertus, J. D., Wright, H. T. and Xuong, Ng. H. (1970), *Biochemistry*, **9**, 1997–2009.

[23] Wright, H. T. (1973), *J. Mol. Biol.*, **79**, 1–11 and 13–23.

[24] Kerr, M. A., Walsh, K. A. and Neurath, H. (1976), *Biochemistry*, **15**, 5566–5570.

[25] Stroud, R. M., Kay, L. M. and Dickerson, R. E. (1974), *J. Mol. Biol.*, **83**, 185–208.

[26] Bode, W. (1979), *J. Mol. Biol.*, **197**, 357–374.

[27] Kerr, M. A., Walsh, K. A. and Neurath, H. (1975), *Biochemistry*, **14**, 5088–5094.

[28] Titani, K., Ericsson, L. H., Walsh, K. A. and Neurath, H. (1975), *Proc. Nat. Acad. Sci.*, **72**, 1666–1670.

[29] Bayliss, W. M. and Starling, E. H. (1902), *J. Physiol.*, **28**, 325–353.

[30] Warren, G. (1982), *Nature*, **297**, 624–625.

[31] Meyer, D. I., Krause, E. and Dobberstein, B. (1982), *Nature*, **297**, 647–650.

[32] Davie, E. W., Fujikawa, K., Kurachi, K. and Kiesel, W. (1977), *Adv. Enzymol.*, **48**, 277–318.

[33] Jackson, C. M. and Nemerson, Y. (1980), *Ann. Rev. Biochem.*, **49**, 765–811.

[34] Kisiel, W., Fujikawa, K. and Davie, E. W. (1977), *Biochemistry*, **16**, 4189–4194.

[35] McKusick, V. A. (1965), *Scientific American*, **213**, 88–95.

[36] Bloom, A. L. and Thomas, O. P. (eds), (1981), *Haemostatis and Thrombosis*, Churchill-Livingstone, Edinburgh.

[37] Cochrane, C. G. and Griffin, J. H. (1982), *Advances in Immunology*, **33**, 241–306.

[38] Ratnoff, O. D. (1973), in *The Molecular Basis of Inherited Disease*, (eds J. B. Stanbury, J. B. Wyngaarden and D. S. Frederickson), McGraw-Hill, New York, pp. 1670–1709.

[39] Magnusson, F. (1971), in *The Enzymes* III, (3rd Edition), (ed. P. D. Boyer), Academic Press, London, pp. 277–321.

[40] Titani, K., Fujikawa, K., Enfield, D. L., Ericsson, L. H., Walsh, K. A. and Neurath, H. (1975), *Proc. Nat. Acad. Sci.*, **72**, 3082–3086.

[41] Katayami, K., Ericsson, L. H., Enfield, D. L., Walsh, K. A., Neurath, H., Davie, E. W. and Titani, K. (1979), *Proc. Nat. Acad. Sci.* (USA), **76**, 4990–4994.

[42] Fujikawa, K., Coan, M. H., Legaz, M. E. and Davie, E. W. (1974), *Biochemistry*, **13**, 5290–5299.

[43] Sottrup-Jensen, L., Zajdel, M., Claeys, H., Petersen, T. E. and Magnusson, S. (1975), *Proc. Nat. Acad. Sci.* (USA), **72**, 2577–2581.

[44] Suttie, J. W. (1980), *Critical Reviews of Biochemistry*, **8**, 191–223, Chemical Rubber Company.

[45] Stenflo, J. (1974), *J. Biol. Chem.*, **249**, 5527–5535.

[46] Stenflo, J., (1976), *J. Biol. Chem.*, **251**, 355–360.

[47] Discipio, R. G., Hermodson, M. A., Yates, S. G. and Davie, E. W. (1977), *Biochemistry*, **16**, 698–702.

[48] Prowse, C. V. and Esnouf, M. P. (1977), *Biochem. Soc. Trans.*, **5**, 255.

[49] Kisiel, W. and Davie, E. W. (1981), *Methods Enzymol.*, **80**, 320–332.

[50] Muller-Eberhard, H. J. (1975), *Ann. Rev. Biochem.*, **44**, 697–724.

[51] Muller-Eberhard, H. J. (1978), in *Molecular Basis of Biological Degradation Processes*, (eds R. D. Berliner, H. Herman, I. H. Lepow and J. M. Tanzer) Academic Press, New York, pp. 65–114.

[52] Porter, R. R. and Reid, K. B. M. (1979), *Adv. Prot. Chem.*, **33**, 1–71.

[53] Reid, K. B. M. and Porter, R. R. (1981), *Ann. Rev. Biochem.*, **50**, 433–464.

[54] Arlaud, G. J., Gagnon, J. and Porter, R. R. (1982), *Biochem. J.*, **201**, 49–59.

[55] Carter, P. E., Dunbar, B. and Fothergill, J. E. (1982), *Biochem. Soc. Trans.*, **10**, 441–442.

[56] Kerr, M. A. (1981), *Biochem. J.*, **189**, 173–181.

[57] Kerr, M. A. (1979), *Biochem. J.*, **183**, 615–622.

[58] Crossley, L. G. (1981), in *Methods in Enzymology*, Vol. 80, Academic Press, London and New York, pp. 112–124.

[59] Nussenzweig, V. and Melton, R. (1981), in *Methods in Enzymology*, Vol. 80, Academic Press, London and New York, pp. 124–133.

[60] Ondetti, M. A. and Cushman, D. W. (1982), *Ann. Rev. Biochem.*, **51**, 283–308.

[61] Docherty, K. and Steiner, D. F. (1982), *Ann. Rev. Physiol.*, **44**, 625–638.

[62] Schapiro, B. M., Schackmann, R. W. and Gabel, C. A. (1981), *Ann. Rev. Biochem.*, **50**, 815–843.

[63] Bornstein, P. and Sage, H. (1980), *Ann. Rev. Biochem.*, **49**, 957–1003.

[64] De Haas, G. H. Slotboom, A. J., Bonsen, P. P. N., Van Deenen, L. L. H., Maroux, S., Puigserver, A. and Desnuelle, P. (1970), *Biochem. Biophys. Acta*, **221**, 31–53.

[65] Cabib, E., Roberts, R. and Bowers, B. (1982), *Ann. Rev. Biochem.*, **51**, 763–794.

4 Enzyme regulation by reversible phosphorylation: the neural and hormonal control of cellular activity

The presence of phosphorus in proteins has been known for over 100 years, but its importance has only been realized since the discovery of enzyme regulation by reversible phosphorylation. The current excitement in this phenomenon stems from the work of Edwin Krebs, Edmond Fischer and Joseph Larner over the period 1955–1970, when they discovered that the neural and hormonal control of glycogen metabolism was mediated by changes in the phosphorylation state of glycogen phosphorylase [1], phosphorylase kinase [2] and glycogen synthase [3].

These three enzymes remained the only examples of this phenomenon until the late 1960s, but the situation changed rapidly following the discovery of cyclic AMP dependent protein kinase [4]. The past 10 years have seen an extraordinary and still accelerating growth in this area. About 40 enzymes are known to be regulated in this manner, and protein phosphorylation is recognized as the major mechanism by which intracellular events are controlled by extracellular (neural and hormonal) stimuli. The first part of this chapter will discuss the control of glycogen metabolism in skeletal muscle since this system is understood in the greatest molecular detail, and serves as a model with which others can be compared. The second part will discuss the regulation of other metabolic pathways and the role of protein phosphorylation in coordinating intracellular responses to neural and hormonal signals.

4.1 Glycogen metabolism in skeletal muscle

Glycogen is the principal storage form of carbohydrate in mammalian cells, and in skeletal muscle, its conversion to lactic acid by anaerobic glycolysis provides much of the ATP required to sustain muscle contraction. It is therefore essential that the rate of glycogenolysis be closely synchronized with the onset, strength and duration of contraction. Glycogen can also be mobilized in *resting muscle* in response to adrenalin, a hormone released from the adrenal gland during periods of stress. This provides a mechanism for mobilizing energy reserves in advance of contraction, thereby anticipating an increased energy demand.

Glycogen is resynthesized in resting muscle, and this process is accelerated by insulin, a hormone secreted by the β-cells of the pancreas in response to elevated blood glucose concentrations. Insulin promotes the transport of glucose into muscle by activating a glucose transporter in the plasma membrane [5], and diverts much of the glucose taken up by this tissue into glycogen synthesis.

4.2 Discovery of enzyme regulation by reversible phosphorylation

The rate limiting enzyme in glycogenolysis is glycogen phosphorylase which catalyses sequential phosphorolysis of the α-1,4 linked glucosyl units in glycogen to produce glucose-1-phosphate (G1P). As long ago as 1938, phosphorylase was isolated in two forms termed *b* and *a* [6]. Phosphorylase *b* was totally dependent on adenosine 5′ monophosphate (AMP) for activity, whereas phosphorylase *a* was almost fully active without AMP. It was reasoned that phosphorylase *a* must contain tightly bound AMP, and that a further enzyme detected in muscle at this time, which converted phosphorylase *a* to *b* (termed prosthetic group removing, or PR, enzyme) removed the bound AMP [7]. However, this could never be confirmed, and over ten years elapsed before the real nature of the reaction was discovered. In 1956, Krebs and Fischer showed that phosphorylase *b* could be converted to phosphorylase *a* by a Mg-ATP requiring enzyme termed phosphorylase kinase [1], which catalyses transfer of the γ-phosphoryl group of ATP to a unique serine hydroxyl located close to the N-terminus of the protein [8]. The activity that reconverted phosphorylase *a* to *b* was therefore a phosphate releasing (or PR!) enzyme, nowadays termed protein phosphatase-1 [9].

4.3 The control of muscle contraction [10]

It is first necessary to outline current ideas about this process before discussing the mechanisms for synchronizing glycogenolysis and contraction.

Skeletal muscle fibres consist of regular arrays of two types of filament termed A and I (Fig. 4.1) Muscle contraction involves a process by which I-filaments (composed mainly of the protein actin) 'slide' passed A-filaments (composed mainly of the protein myosin). This leads to a shortening of the total length of the fibre (although the volume remains constant) and tension is generated.

Mammalian skeletal muscle myosins are composed of a large polypeptide termed the *'heavy'* chain ($M_r \cong 200\,000$) and two types of *'light'* chain ($M_r \cong 20\,000$). Each molecule contains two copies of each of the three subunits. Under the electron microscope, each myosin molecule has the appearance of a double helical rod, which terminates in two globular heads, termed a cross bridge (Fig. 4.2a). The *'light'* chains are present exclusively

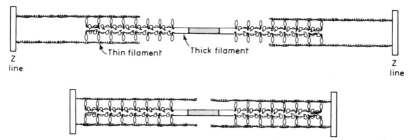

Fig. 4.1 Schematic representation of the arrangement of the thick (A) and thin (I) filaments in relaxed (upper) and contracting (lower) muscle. The diagram shows a longitudinal view of a sarcomere, the repeating unit of the myofibril. There are many thousands of myofibrils in a muscle fibre.

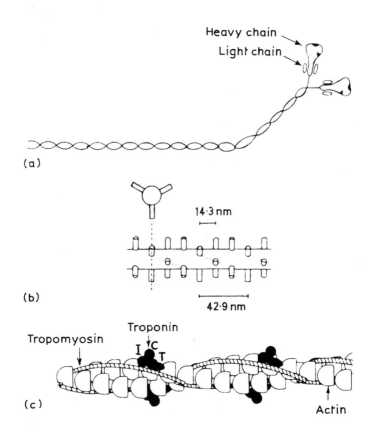

(a)

14·3 nm

(b)

42·9 nm

Tropomyosin

Troponin

I C

T

(c)

Actin

Fig. 4.2 Schematic representation of an A and an I filament. (a) single myosin molecule; (b) a thick (A) filament in which the myosin cross-bridges (projections) lie on a 3-stranded helix; (c) a thin (I) filament: one troponin molecule is associated with every seventh actin monomer. Each troponin molecule contains equimolar amounts of TN-C, TN-I and TN-T [10].

in the head regions. The myosin molecules are packed together in the A-filament in an ordered helical manner, with the cross bridges projecting (Fig. 4.2b). The cross bridges possess ATPase activity, so that myosin is in fact an enzyme. The sliding of actin relative to myosin during contraction involves an interaction of actin with the head region of myosin, and is accompanied by 100-fold activation or more of actomyosin ATPase. The force generating mechanism is a consequence of the linkage of actin and myosin, while the hydrolysis of ATP provides the energy for the performance of mechanical work.

The actin monomer is a globular protein (M_r = 42 000), which polymerizes as a double helical array in the I-filament. The I-filament also contains two further proteins, troponin and tropomyosin (Fig. 4.2c). Troponin is a complex of three subunits, TNT, TN-I and TN-C. Tropomyosin and TN-I are the components that prevent interaction of actin and myosin, and hence the activation of myosin ATPase, in resting muscle. At rest, virtually all the calcium ions in the muscle cell are retained in a membrane system surrounding the myofibril, termed the sarcoplasmic reticulum, by virtue of an active transport mechanism. Following stimulation of the nerve to a muscle, the sarcoplasmic reticulum becomes permeable to Ca^{2+} by a mechanism which is not yet understood, and the concentration of Ca^{2+} in the muscle cytoplasm rises from 0.1 μM (or below) to about 10 μM. Ca^{2+} binds to TN-C causing it to bind more strongly to TN-I. This neutralizes the inhibition of actomyosin ATPase by tropomyosin-TN-I, ATP is hydrolysed and contraction initiated. TN-T interacts with tropomyosin, and is necessary to restore full sensitivity to Ca^{2+} in reconstituted systems. When nervous stimulation ceases, Ca^{2+} is rapidly withdrawn into the sarcoplasmic reticulum and contraction ceases.

The contractile apparatus can therefore be thought of as a rather complex enzyme. There is a catalytic subunit (the 'cross bridge' portion of the heavy chain) and many regulatory subunits, one of which, TN-C, binds the allosteric activator Ca^{2+}. The binding of Ca^{2+} triggers a series of allosteric transitions, which lead to activation of the ATPase, and a 20–50% shortening of the length of the sarcomere. The concentration of Ca^{2+} is itself controlled by nerve impulses acting through the sarcoplasm reticulum.

4.4 Synchronization of glycogenolysis and muscle contraction: activation of phosphorylase kinase by Ca^{2+}, calmodulin and troponin

Since actomyosin ATPase activity can increase more than 100-fold during the few milliseconds needed for maximum tension to develop, and ATP in resting muscle (about 7 mM) is sufficient to support vigorous contraction for less than 1 second, it is essential that effective mechanisms exist for regenerating ATP as it is hydrolysed. This can be provided for a further few seconds by the store of creatine phosphate (25 mM), but beyond this time, regeneration of ATP from ADP depends almost exclusively on the conversion of glycogen to lactate.

The rate of glycogenolysis can also increase several hundred fold within seconds of the onset of contraction, implying a corresponding increase in the activity of glycogen phosphorylase. It is now established that phosphorylase is almost entirely in the b-form in resting muscle. When isolated frog muscles are made to contract, by giving shocks of fixed duration and frequency, conversion to the a-form is promoted. At any frequency of stimulation a steady state level of phosphorylase a is reached which reflects the relative activities of phosphorylase kinase and protein phosphatase-1 (Fig. 4.3). The level of phosphorylase a that is reached increases and the time taken to reach the steady state decreases, with increasing frequency of stimulation. At frequencies which cause the muscle to go into a tetanus, a steady state level of 70% phosphorylase a is reached with a half time of less than 1 second (Fig. 4.4). The rate of glycogenolysis also increases with increasing frequency of stimulation, reaching over 100-times the basal rate at only 0.8 shocks per second when the steady state level of phosphorylase a is about 10% (Fig. 4.3) [13].

These experiments suggested that the appearance of phosphorylase a was an important factor in synchronizing glycogenolysis with muscle

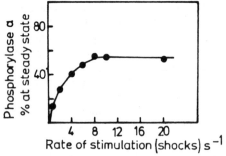

Fig. 4.3 Effect of electrical stimulation on the level of phosphorylase *a* in isolated frog sartorius muscle. The half-time required to reach the steady state was 30 seconds at two shocks per second, and 10 seconds at six shocks per second [11].

contraction. They also indicated that phosphorylase kinase must be inactive in resting muscle and converted to an active form upon contraction. It is now clear that the signal which activates phosphorylase kinase is the same as that which initiates contraction, i.e. calcium ions (Fig. 4.5).

Phosphorylase kinase is composed of four types of subunit α, β, γ, δ and has the structure $(\alpha\beta\gamma\delta)_4$ [15]. The γ-subunit is catalytically active [16] and the δ-subunit is the Ca^{2+}-binding component [17]. The binding of Ca^{2+} to the δ-subunit is a prerequisite for activation of the γ-subunit.

The δ-subunit is identical to a protein termed calmodulin (Fig. 4.6). Calmodulin is closely related in structure to TN-C, the protein that confers Ca^{2+}-sensitivity to the muscle contractile apparatus (Section 4.3). Like TN-C it can bind up to four calcium ions per mol at micromolar concentrations, the affinity of the two Ca^{2+}-binding sites in the N-terminal half of the molecule being 10-fold higher than those in the C-terminal domain [19].

Phosphorylase kinase has been found to interact with a second molecule of calmodulin, termed the δ'-subunit, which activates the dephosphorylated form of the enzyme considerably [14]. The δ'-subunit only binds to phosphorylase kinase in the presence of Ca^{2+}, where it interacts with the α and β-subunits. This contrasts with the δ-subunit which remains complexed with the γ-subunit even in the absence of Ca^{2+} [20]. TN-C, the troponin

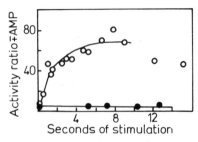

Fig. 4.4 Effect of tetanic contraction on the level of phosphorylase *a* in mouse caudofemoralis muscle *in situ*: ○—normal mice (strain C-57); ●—abnormal mice (strain I) which completely lack phosphorylase kinase [12].

46

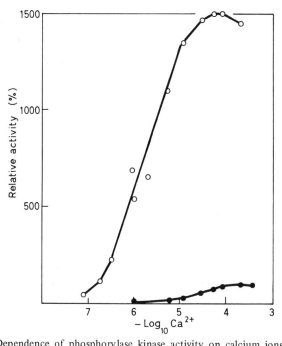

Fig. 4.5 Dependence of phosphorylase kinase activity on calcium ions: ●—dephosphorylated b-form; ○—phosphorylated a-form. The b-form (but not the a-form) can be activated 20–30 fold by TN-C at 1–3 μM Ca^{2+} [14].

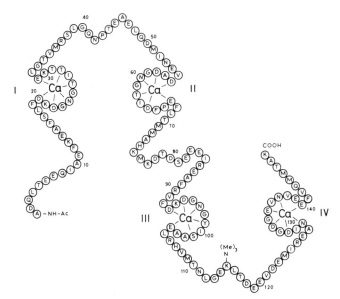

Fig. 4.6 Amino acid sequence of calmodulin, utilizing the one letter code for amino-acids. The four proposed Ca^{2+} binding domains are indicated [18].

complex, and even artificial thin filaments of muscle can substitute for the δ'-subunit in the activation of phosphorylase kinase *in vitro*, and TN-C rather than the δ'subunit may be the important activator *in vivo* [14,15]. Thus, while interaction of Ca^{2+} with the δ-subunit is essential for activation, as much as 20–30 fold higher activation can be achieved through the interaction with TN-C at micromolar concentrations of Ca^{2+}. Activation of glycogenolysis and muscle contraction by the same CA^{2+}-binding protein (TN-C) would appear to represent an attractive mechanism for coupling these two processes (Fig. 4.7).

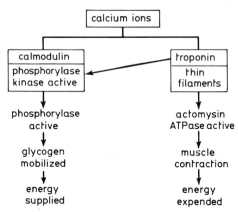

Fig. 4.7 Role of Ca^{2+}, calmodulin and troponin-C in synchronizing glycogenolysis and muscle contraction.

4.5 The role of substrates and allosteric effectors in the regulation of glycogenolysis

The regulation of phosphorylase *b* by AMP was the first example of the control of enzyme activity by an allosteric effector [6]. Phosphorylase *b* is inactive in the absence of AMP, but is just as active as phosphorylase *a* in its presence. The regulation of phosphorylase *b* by AMP is however critically dependent on the substrate inorganic phosphate (P_i). As the concentration of P_i increases, the concentration of AMP required for half maximal activation (K_a) decreases, and conversely, the K_m for P_i decreases as AMP increases [21]. P_i and AMP are therefore *synergistic* activators of phosphorylase. On the other hand, activation by AMP is antagonized by both ATP and glucose-6-phosphate (G6P).

Phosphorylase *a* is almost fully active in the absence of AMP, but only at saturating substrate concentrations. At low levels of P_i and glycogen, the *a*-form is also largely dependent on AMP [22]. Moreover the K_a for AMP of the *a*-form is about 100-fold lower than the *b*-form. ATP and G6P are not inhibitors of phosphorylase *a*.

The potential importance of AMP, P_i and ATP in the control of glycogenolysis can be seen by inspection of Fig. 4.8 which shows that the steady state concentrations of these compounds are a function of eight reactions. Muscle contraction (reaction 1) tends to decrease ATP and increase P_i,

$$\text{ATP} \longrightarrow \text{ADP} + P_i \tag{1}$$

$$\text{creatine-P} + \text{ADP} \rightleftharpoons \text{creatine} + \text{ATP} \tag{2}$$

$$2\,\text{ADP} \rightleftharpoons \text{AMP} + \text{ATP} \tag{3}$$

$$\text{glycogen}_n + P_i \longrightarrow \text{glycogen}_{n-1} + \text{G1P} \tag{4}$$

$$\text{G3P} + \text{NAD} + P_i \rightleftharpoons 1,3\,\text{DPG} + \text{NADH} + \text{H}^+ \tag{5}$$

$$\text{AMP} \longrightarrow \text{IMP} + \text{NH}_3 \tag{6}$$

$$\text{aspartate} + \text{GTP} + \text{IMP} \longrightarrow \text{adenylosuccinate} + \text{GDP} + P_i \tag{7}$$

$$\text{adenylosuccinate} \longrightarrow \text{AMP} + \text{fumarate} \tag{8}$$

Fig. 4.8 Reactions determining the levels of ATP, AMP and P_i in muscle. Enzymes: 1–actomyosin ATPase; 2–creatine kinase; 3–adenylate kinase; 4–phosphorylase; 5–glyceraldehyde-3-phosphate (G3P) dehydrogenase; 6–AMP deaminase; 7–adenylosuccinate synthetase; 8–adenylosuccinase.

although reaction 2 will tend to buffer ATP levels. Glycogenolysis will not only regenerate ATP but consume P_i through reactions 4 and 5. AMP should be inversely related to ATP through reactions 1 and 3. Increased levels of P_i and AMP are therefore signals of increased ATP hydrolysis and decreasing ATP levels respectively, and hence appropriate activators of glycogenolysis whose primary function is to replenish ATP. Similarly, ATP and G6P are appropriate inhibitors of glycogenolysis, the former metabolite acting as a classical feedback inhibitor, and the latter as a signal of the activity of phosphofructokinase (PFK), the rate determining enzyme in the glycolytic pathway. However these arguments are only theoretical, and do not prove that these effectors play a regulatory role *in vivo*.

The most direct way to examine the role of AMP, P_i, ATP and G6P would be to measure the concentrations of these metabolites at different states of metabolic activity. Unfortunately these measurements are beset with many difficulties and pitfalls, that are discussed in another volume of this series [23]. A major problem is that classical methods only measure the average metabolite concentrations in a whole muscle. Since each muscle is composed of thousands of muscle fibres and several fibre types, and each muscle cell contains many subcellular organelles, the values obtained may be rather gross averages which bear little resemblance to the concentrations available to phosphorylase which is located in the muscle cytoplasm. Furthermore, classical techniques only measure the total amount of a metabolite, whereas it is essential to evaluate the 'free' (i.e. metabolically available) concentration. This may differ considerably from the total concentration as a result of tight binding to specific macromolecules. For example, most of the ADP in skeletal muscle is bound to actin.

Recently, it has become possible to estimate the concentrations of certain metabolites in *intact* tissues by the use of nuclear magnetic resonance techniques [24,25]. These investigations have already suggested that the 'free' concentrations of P_i, ADP and AMP in resting muscles may be much lower than those determined by classical methods, and explains why phosphorylase *b* is inactive in resting muscle ($< 0.1\%$ of maximal potential activity).

Nevertheless, two lines of evidence indicate that changes in substrate and/or effector concentrations *are* important in activating glycogenolysis during contraction. When resting frog muscles are perfused with adrenalin (Section 4.6), the level of phosphorylase a rises to 50%, but glycogenolysis is only accelerated 10-fold, indicating that phosphorylase a is operating at only approximately 1% of its maximal activity. In contrast, glycogenolysis is stimulated 100-fold when the muscle is stimulated electrically at frequencies that increase the level of phosphorylase a to only 10% (Fig. 4.3), [12,26,27]. The simplest explanation for this discrepancy is that the P_i + AMP/ATP + G6P ratio is much higher in contracting than in resting muscle. Since phosphorylase a binds AMP with much higher affinity than phosphorylase b and the concentration of phosphorylase in skeletal muscle (0.08 mM) is higher than that of AMP, this nucleotide should activate the a-form preferentially *in vivo*.

The second experiment that bears on this problem, has been carried out with I-strain mice that lack phosphorylase kinase, and cannot convert phosphorylase b to a (Fig. 4.4). Stimulation at frequencies producing a tetanus does not initiate glycogenolysis immediately, unlike normal mice. However, after a lag of a few seconds, glycogenolysis does take place and can approach 50% of the maximal rate seen with normal mice [13]. The following inferences can be drawn from these observations.

(a) Phosphorylase b can be activated *in vivo* (presumably by AMP) if conversion to phosphorylase a fails to occur.

(b) Significant increases in AMP and/or P_i must occur within seconds of contraction in I-strain mice.

(c) The lag before glycogenolysis is initiated in I-strain mice, might represent the time required for reactions 1–3 (Fig. 4.8) to raise AMP and P_i to levels that can activate phosphorylase b.

(d) The absence of a lag in normal mice strongly supports the idea that the appearance of phosphorylase a initiates glycogenolysis in normal skeletal muscle, and that phosphorylase a is activated by a lower AMP + Pi/ATP + G6P ratio than the b-form *in vivo*.

I-strain mice appear relatively normal, and are able for example, to swim for as long as normal mice. In the controlled conditions of the laboratory, the ability to degrade glycogen immediately, and at a slightly faster rate is not particularly important, but must have a strong selective advantage in the wild, for catching prey and escaping from predators.

AMP may play a further role, since it inhibits the reconversion of phosphorylase a to phosphorylase b *in vitro* (K_i 0.01 mM). This inhibition is caused by the binding of AMP to phosphorylase a and not protein phosphatase-1 [28]. This is a form of regulation not previously encountered in this book, where an allosteric effector inhibits an enzyme by altering the conformation of its *substrate*. AMP may therefore not only activate phosphorylase directly but also influence the steady state level of phosphorylase a. This type of control is observed with a number of enzymes that are regulated by reversible phosphorylation, and may be important in allowing allosteric effectors to exert their effects at two different levels.

4.6 The control of glycogenolysis by adrenalin

4.6.1 Discovery of cyclic AMP [29]

Although the activation of glycogenolysis by adrenalin in skeletal muscle and liver was documented in the 1920s, little or no progress was made in understanding the mechanism of action of this (or any other) hormone until 1950. The reason was a complete inability to demonstrate the effect of a hormone in any system other than the intact animal. However in the early 1950s Earl Sutherland showed an enhancement of glycogenolysis and phosphorylase activity by incubating liver slices with adrenalin, the first report that a hormone could influence the activity of a specific enzyme. However, the response to adrenalin was lost when the liver slices were homogenized. With the discovery of the chemical nature of the change in phosphorylase activity (i.e. phosphorylation) it became clear that ATP and Mg^{2+} would be necessary for activation, and indeed addition of these components restored the sensitivity to adrenalin. The ability to demonstrate a hormonal effect reproducibly in a broken cell preparation was a major breakthrough, and allowed the response to be separated into two stages. In the first, adrenalin acted on a membrane fraction to produce a small heat stable factor; in the second, this factor could replace the hormone and increase the proportion of phosphorylase a.

The factor was identified as adenosine $3'5'$cyclic monophosphate (cyclic AMP) (Fig. 4.9). It was shown to be found by an enzyme, termed adenylate cyclase, which converts ATP to cyclic AMP and pyrophosphate, and hydrolysed to AMP by one or more cyclic AMP phosphodiesterases.

The activation of adenylate cyclase by adrenalin requires at least three proteins, all located in the plasma membrane that surrounds each muscle or liver cell [30]. There is a β-receptor (R) that binds adrenalin, a regulatory subunit that binds guanine nucleotides (G_s), and the catalytic subunit of adenylate cyclase itself (C). This three component model is shown in Fig. 4.10. Adenylate cyclase by itself is essentially inactive, but can be stimulated by G_s (s for stimulatory) provided that this protein binds GTP.

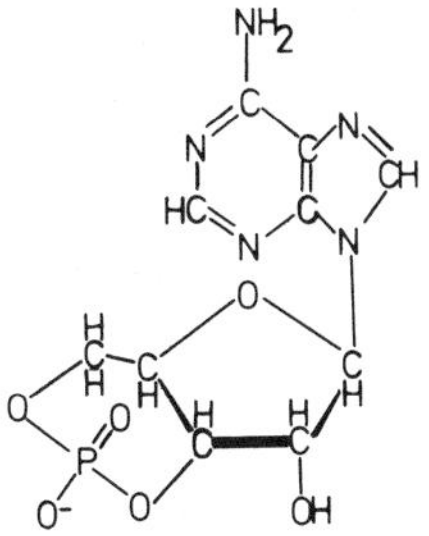

Fig. 4.9 The structure of cyclic AMP.

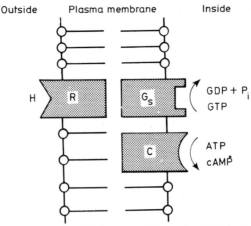

Fig. 4.10 Three-component model for adenylate cyclase (31). H, hormone e.g. adrenalin; R, hormone receptor; Gs, guanine nucleotide binding protein; C, catalytic subunit of adenylate cyclase; cAMP, cyclic AMP.

However, when complexed with C, G_s is a GTPase that hydrolyses GTP to $GDP + P_i$, and this terminates the activation process since G_s-GDP is unable to stimulate adenylate cyclase. It is envisaged that the adrenalin–β-receptor complex activates adenylate cyclase by facilitating the regeneration of G_s-GTP from G_s-GDP. The adenylate cyclase system therefore resembles an allosteric enzyme in which the β-receptor and G_s protein are regulatory subunits and adrenalin and GTP allosteric activators, except that it is arranged with a fixed orientation across the plasma membrane, so that the interaction of adrenalin with the β-receptor on the outer surface, leads to the formation of cyclic AMP intracellularly. It should, however, be emphasized that Fig. 4.10 is only a schematic representation. It is already clear that the R, G_s and C proteins are not present in equimolar proportions and that G_s is present in a large molar excess over both R and C. This implies that one molecule of β-receptor can activate a number of molecules of G_s protein [32]. Further details of the hormone–adenylate cyclase system are discussed in the following chapter (Section 5.4).

4.6.2· *The action of cyclic AMP in glycogenolysis*
The discovery that cyclic AMP promoted the conversion of phosphorylase *b* to *a* indicated that this molecule must activate phosphorylase kinase or inhibit protein phosphatase-1, and current evidence suggests that both mechanisms are important in the regulation of glycogenolysis *in vivo*. Cyclic AMP transmits the hormonal signal by activating an enzyme termed cyclic AMP dependent protein kinase (cAMP-PK) [4]. This is composed of a regulatory subunit (R) which binds two molecules of cyclic AMP, and a catalytic subunit (C), that carries the active site [33]. Activation by cyclic AMP is accompanied by the dissociation of the R_2C_2 complex:

$$R_2C_2 + 4\,cAMP \rightleftharpoons 2C + (cAMP)_4R_2$$

$$\text{(inactive)} \qquad\qquad \text{(active)}$$

Phosphorylase kinase can exist as a low activity dephosphorylated b-form that can be activated 20–30 fold by TN-C at micromolar concentrations of Ca^{2+}, or as a high activity phosphorylated a-form which is essentially unaffected by TN-C [14,15]. The conversion of phosphorylase kinase b to a is catalysed by cAMP-PK, a reaction which involves the rapid phosphorylation of one serine residue on the β-subunit and slower phosphorylation of a further serine on the α-subunit. Phosphorylation of the β-subunit is largely responsible for activation [34] although phosphorylation of the α-subunit may also cause a small additional stimulation of activity [35]. Both serine residues are phosphorylated *in vivo* in response to adrenalin [34].

The b-form and a-form both have an absolute requirement for Ca^{2+}, but the a-form has a 15-fold higher activity at saturating Ca^{2+} and a 10-fold lower K_a for Ca^{2+} (Fig. 4.5). This suggests that phosphorylation of the α- and β-subunits not only increases the catalytic activity of the γ-subunit, but may allow activation to occur when fewer calcium ions are bound to the δ-subunit (calmodulin). It has been suggested that the a-form is fully activated when only the two high affinity Ca^{2+}-binding sites are occupied by Ca^{2+}, whereas all four Ca^{2+}-binding sites must be saturated to activate the b-form (Fig. 4.6).

Protein phosphatase-1 (PrP-1) is not only the major phosphatase that reconverts phosphorylase a to b, but is also the major enzyme that dephosphorylates the β-subunit of phosphorylase kinase and glycogen synthase [36]. It therefore catalyses each of the dephosphorylation reactions that inhibit glycogenolysis and activate glycogen synthesis (Fig. 4.11). PrP-1 is potently inhibited by a protein, termed inhibitor-1, whose activity is only expressed after it has been phosphorylated on a threonine residue by cAMP-PK [37] (Fig. 4.12). Inhibition of PrP-1 by this mechanism is likely to be of considerable importance in the adrenergic control of glycogen metabolism [38]. The regulation of PrP-1 by inhibitor-1 may explain how high levels of phosphorylase a (70%) can be attained in response to adrenalin, even in resting muscle where the Ca^{2+}-concentration is less than or equal to 0.1 μM and phosphorylase kinase a should exhibit only a few percent of its potential activity (Fig. 4.5).

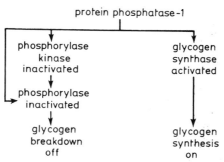

Fig. 4.11 Role of protein phosphatase-1 in the control of glycogen metabolism.

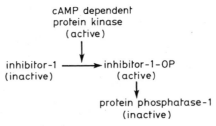

Fig. 4.12 Cyclic AMP dependent protein kinase can inactivate protein phosphatase-1 through the phosphorylation of inhibitor-1.

4.7 Regulation of glycogen synthesis

The rate limiting reaction in this pathway is the transfer of glucosyl residues to glycogen from uridine diphosphate glucose (UDPG), catalysed by glycogen synthase. Although this enzyme was shown to be regulated by a phosphorylation–dephosphorylation mechanism in the early 1960s [3] it is only quite recently that the complexity of this system has been fully appreciated. Glycogen synthase is phosphorylated on seven serine residues by at least six protein kinases, and all seven sites are phosphorylated *in vivo* [39–42]. Since the enzyme is composed of four identical subunits (M_r = 86 000), the number of possible phosphorylated species is exceedingly large. The organization of the phosphorylation sites in the polypeptide chain is shown in Fig. 4.13. Cyclic AMP-PK phosphorylates sites 1a, 1b and 2, while phosphorylase kinase, a calmodulin dependent glycogen synthase kinase and glycogen synthase kinase-4 (GSK-4) phosphorylate site 2. Glycogen synthase kinase-3 (GSK-3) labels sites (3a + 3b + 3c) and glycogen synthase kinase-5 (GSK-5) site 5.

The effect of phosphorylation is to decrease the activity of glycogen synthase, but the changes in kinetic properties are complex, and different

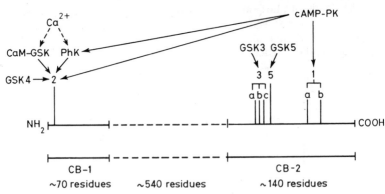

Fig. 4.13 Organization of the phosphorylation sites in rabbit skeletal muscle glycogen synthase. Site 2 is seven residues from the N-terminus, and sites 3a, 3b, 3c, 5, 1a and 1b are 30, 34, 38, 46, 87 and 100 residues from the N-terminus of a large cyanogen bromide peptide (CB-2) at the C-terminal end of the molecule [40]. Abbreviations: cAMP-PK, cyclic AMP dependent protein kinase; PhK, phosphorylase kinase; GSK, glycogen synthase kinase; CaM, calmodulin.

phosphorylation sites have different effects. The K_m for UDPG increases progressively as the extent of phosphorylation increases [43]. Phosphorylation of sites (3a + 3b + 3c) increases the K_m for UDPG to a greater extent than site 2 or site 1a, but the effects are additive so that even larger changes are produced when all five sites are phosphorylated. The effects of phosphorylation can be reversed by the allosteric activator G6P which decreases the K_m for UDPG. However, phosphorylation also increases the K_a for G6P and decreases the K_i for inhibitors (e.g. P_i) that antagonize activation by G6P.

Neither phosphorylation of site 1b nor site 5 appear to alter the kinetic properties of glycogen synthase significantly.

4.7.1 Regulation of glycogen synthesis during muscle contraction

Glycogen synthase can be phosphorylated at site-2 by two Ca^{2+}-calmodulin dependent protein kinases. One of these is phosphorylase kinase [44] while the other is a distinct enzyme that does not convert phosphorylase b to a [42]. These findings suggest that calcium ions not only activate glycogenolysis, but also inhibit glycogen synthesis. At physiological concentrations of phosphorylase ($\cong 80\,\mu M$) and glycogen synthase ($\cong 3\,\mu M$) phosphorylase kinase phosphorylates the former enzyme preferentially. It has therefore been suggested that, during contraction, the Ca^{2+}-calmodulin dependent glycogen synthase kinase might initially phosphorylate site 2, phosphorylase kinase only acting on this site if the conversion of phosphorylase b to a is essentially complete [42].

However, it has not yet been demonstrated that increased phosphorylation of site 2 occurs *in vivo* during contraction. Furthermore, increased phosphorylation at site 2 may be counterbalanced by increased rates of dephosphorylation resulting from glycogen depletion (Section 4.7.4).

4.7.2 Inhibition of glycogen synthesis by adrenalin [41]

The phosphate content of glycogen synthase increases from slightly below 3 mol per subunit in the absence of adrenalin to slightly above 5 mol per subunit, in the presence of maximally effective doses of this hormone. The phosphorylation of sites 1a and 1b each increase by approx 0.25 mol per subunit, site 2 by approx 0.6 mol per subunit and sites (3a + 3b + 3c) by approx 1.2 mol per subunit. Inhibition of glycogen synthase by adrenalin is therefore largely the result of increased phosphorylation at sites (3a + 3b + 3c), with a smaller contribution from sites 2 and 1a.

The effects of adrenalin are mediated by cyclic AMP, yet the activity of GSK-3 is unaffected by this molecule. The increased phosphorylation at sites (3a + 3b + 3c) must therefore be caused by an inhibition of protein phosphatase-1. One mechanism by which this could take place is through the phosphorylation of inhibitor-1 (Fig. 4.12). However, the high levels of phosphorylase a produced in response to adrenalin may also decrease the rate of dephosphorylation of glycogen synthase by simple competition for protein phosphatase-1. In view of the high concentration of phosphorylase (80 μM) in skeletal muscle relative to glycogen synthase (3 μM), this additional mechanism may be important.

The phosphorylation of sites 1a and 1b is presumably catalysed by cAMP-PK, while the increased phosphorylation at site 2 may either be catalysed directly by cAMP-PK, indirectly through the activation of phosphorylase kinase by cAMP-PK (Fig. 4.13), by the inhibition of protein phosphatase-1, or by any combination of these mechanisms.

4.7.3 Stimulation of glycogen synthesis by insulin [45]

Insulin stimulates glycogen synthesis in skeletal muscle by promoting the dephosphorylation and activation of glycogen synthase, an effect that is independent of the action of insulin on glucose transport [46]. Activation of glycogen synthase is rapid, occurring within minutes *in vivo*, although it is not as rapid as the action of adrenalin which occurs within seconds. The effect of insulin occurs in the presence of propranolol (which blocks the action of adrenalin), and without any detectable change in the concentration of cyclic AMP.

In the presence of maximally effective doses of insulin, the phosphate content of glycogen synthase decreases by approx 0.5 mol per subunit and results from a specific dephosphorylation of sites (3a + 3b + 3c). Insulin must therefore either decrease the activity of GSK-3 or activate the protein phosphatase(s) that acts on sites (3a + 3b + 3c). The two enzymes in skeletal muscle that are capable of dephosphorylating these sites *in vitro* are protein phosphatase-1 and protein phosphatase-2A, of which the former appears to be the dominant activity [36]. However neither phosphatase is specific for sites (3a + 3b + 3c) *in vitro*, and site 2 is dephosphorylated with almost equal rapidity. This suggests that activation of glycogen synthase by insulin may involve a specific inhibition of GSK-3. Possible mechanisms by which this might take place are discussed in Section 4.11.

4.7.4 Role of G6P and glycogen in the regulation of glycogen synthase

The glucose transported into skeletal muscle under the influence of insulin is immediately converted to G6P, which in resting muscle is mostly used for glycogen synthesis. Thus insulin not only promotes the dephosphorylation of glycogen synthase causing a decrease in the K_a for UDPG, but increases the concentration of G6P which produces a further decrease in the K_m for UDPG. Dephosphorylation also decreases the K_a for G6P, further enhancing the action of this effector. Activation of glycogen synthase by G6P is an example of *feedforward control* where an early metabolite signals to a later enzyme, that an increased activity of the pathway is required.

The phosphorylation state of glycogen synthase in skeletal muscle increases as the glycogen content increases and vice versa [47], and this may represent an important feedback control by which glycogen can regulate the rate and extent to which it is resynthesized. However, the molecular basis for this effect is unclear, since glycogen does not influence the rate of phosphorylation or dephosphorylation of glycogen synthase *in vitro*, whichever protein kinase or protein phosphatase is used for such experiments [45,48].

4.7.5 Role of site 5 [48]

The phosphorylation of site 5 by GSK-5 does not appear to modify the kinetic properties of glycogen synthase, yet this residue is almost fully phosphorylated *in vivo* under all conditions so far examined [41,45], suggesting it has a biological role. It has recently been discovered that phosphorylation of sites (3a + 3b + 3c) by GSK-3 cannot take place unless site 5 is phosphorylated. Thus GSK-5 appears to be the first example of a protein kinase whose function is to form the recognition site for another protein kinase.

4.8 Cyclic AMP and hormone action

The discovery of cyclic AMP immediately solved two of the great mysteries of hormone action. Firstly, it explained how hormones could exert their effects without actually entering the cells whose metabolic activity they regulated. Secondly, the formation of micromolar concentrations of cyclic AMP within the cell in response to nanomolar or subnanomolar increases in hormone levels in the blood started to explain their remarkable biological potency.

The formation of cyclic AMP does not just take place in skeletal muscle or liver in response to adrenalin, but occurs in a variety of tissues in response to the interaction of many different hormones with their receptors. In cells where two or more hormones can stimulate adenylate cyclase, each hormone–receptor complex generally stimulates the same adenylate cyclase by a mechanism which may be identical to that shown in Fig. 4.10 [29]. Cyclic AMP is therefore a general mediator of hormone action, and its identification must rank as one of the great biological discoveries of the century (E. W. Sutherland, Nobel Prize 1971).

Cyclic AMP-PK appears to be the only protein in mammalian cells that binds cyclic AMP with high affinity (apart from cyclic AMP phosphodiesterase), and its activity is similar in all tissues, even those where glycogen metabolism is of very minor importance. The hypothesis therefore arose that the specificity of hormones which act through cyclic AMP, is determined by whether their receptors are present on the plasma membrane of a target cell, and which physiological substrates for cAMP-PK are present within those cells [49,50] (Fig. 4.14).

In the case of substrates that are enzymes, the effect of phosphorylation is often to change the K_m for a substrate, the K_a for an activator or the K_i for an inhibitor, as clearly illustrated by the enzymes of glycogen metabolism. Substrates, activators and inhibitors may also affect the rate at which an enzyme is phosphorylated or dephosphorylated, thereby amplifying or suppressing the effects of covalent modification. The interplay between phosphorylation–dephosphorylation, allosteric effectors and substrates respresents the means by which extracellular (neural and hormonal) and intracellular information is integrated to determine the precise activity of a metabolic pathway *in vivo.* The extent to which phosphorylation will alter the activity of an enzyme *in vivo* therefore depends upon the metabolic state of the cell. Some of the enzymes that are substrates for cAMP-PK are discussed below.

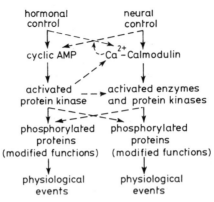

Fig. 4.14 Molecular mechanisms by which cyclic AMP and calmodulin mediate the neural and hormonal control of cellular activity. The broken lines show some of the points at which the cyclic AMP and Ca^{2+}–calmodulin systems can be interlinked [9].

4.8.1 Mobilization of triglycerides and cholesterol esters

Triglycerides stored in adipose tissue are hydrolysed to fatty acids and glycerol in response to the hormones adrenalin and glucagon, allowing this energy reserve to be mobilized during periods of stress or in the fasted state (Section 4.8.2). These effects are mediated by cAMP-PK which catalyses the phosphorylation and activation of triglyceride lipase [51].

Cholesterol esters stored in adrenal cortex are mobilized in response to adrenocorticotrophic hormone (ACTH) to produce cholesterol for the biosynthesis of several steroid hormones, such as cortisol. Cholesterol esters are also hydrolysed in the corpus luteum in response to luteinizing hormone (LH), providing cholesterol for the biosynthesis of the steroid hormones progesterone and oestradiol. Both these processes are initiated by cAMP-PK which catalyses the phosphorylation and activation of cholesterol ester hydrolase [52].

Interestingly, triglyceride lipase in adipose tissue and cholesterol ester hydrolase in adrenal cortex and corpus luteum appear to be one and the same protein [53]. The hormonal response is therefore not only governed by the type of hormone receptor and phosphorylated enzyme present, but in this case by the nature of the enzyme substrate in the target cell.

4.8.2 Regulation of glucose production by glucagon in mammalian liver

Glucagon is a hormone released from the α-cells of the pancreas when the blood glucose concentration is low (i.e. in the fasted state) which *increases* blood glucose levels by stimulating glycogenolysis and gluconeogenesis in the liver. The mobilization of glycogen occurs by a mechanism which is perfectly analogous to the action of adrenalin on glycogenolysis in skeletal muscle. The stimulation of gluconeogenesis is also mediated by cAMP-PK and involves the phosphorylation of 6-phosphofructo-2-kinase (PFK-2), fructose 2:6 bisphosphatase (F26Pase) and pyruvate kinase.

PFK-2 and F26Pase are the enzymes responsible for the formation and destruction of fructose 2:6 bisphosphate (F26P). This compound, which was only discovered in 1980 [54], is a potent allosteric activator of 6-phosphofructo-1-kinase (PFK-1) and inhibitor of fructose 1:6 bisphosphatase (F16Pase). F26P activates PFK1 by decreasing the K_m for fructose-6-phosphate and increasing the K_i for ATP, and inhibits F16Pase by increasing the K_m for fructose 1:6 bisphosphate. In both cases F26P acts synergistically with AMP, which is also an allosteric activator of PFK1 and inhibitor of F16Pase (Fig. 4.15).

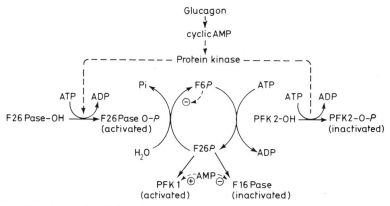

Fig. 4.15 Synthesis and degradation of F26P in the liver, and its regulation by glucagon and metabolites. P-glycerol, glycerol phosphate; F6P, fructose-6-phosphate, (−), inhibition; (+), activation.

Cyclic AMP-PK catalyses the inhibition of PFK2 and activation of F26Pase, causing a rapid drop in the intracellular level of F26P in response to glucagon (Fig. 4.15). The resulting inhibition of PFK1 and activation of F16Pase not only stimulates gluconeogenesis, but also ensures that glycogen is converted by glucose, and not to other compounds via the glycolytic pathway.

The formation and destruction of F26P appears to be catalysed by a single *bifunctional* enzyme containing both PFK2 and F26Pase activity. Inhibition of PFK2 and activation of F26Pase may therefore be triggered by phosphorylation of the same residue [55]. This finding is remarkable but by no means unique. The protein kinase and phosphatase that catalyse inactivation and reactivation of isocitrate dehydrogenase in *E. coli*, an important reaction in the control of the glyoxylate cycle, are both located on the same polypeptide chain [56]. Similarly, a single *bifunctional* protein contains the two enzymes that catalyse the adenylylation and deadenylylation of glutamine synthetase in *E. coli* (Chapter 5). This situation is clearly very useful from a regulatory standpoint since enzymes that catalyse opposing reactions can be controlled in a reciprocal manner either by phosphorylation or by allosteric effectors. For example fructose-6-phosphate, the substrate of PFK-2, inhibits F26Pase (Fig. 4.15). These two enzymes are also controlled by other metabolites [54].

Pyruvate kinase catalyses the last, irreversible reaction of glycolysis, in which pyruvate and ATP are formed from phosphoenolpyruvate (PEP) and

59

ADP. During gluconeogenesis, PEP is reformed from pyruvate, via oxalo-acetate, in two other reactions catalysed by the mitochondrial enzymes pyruvate carboxylase and PEP-carboxykinase. Since the maximal activity of pyruvate kinase is very high compared with those of the gluconeogenic enzymes, it must be carefully controlled. Pyruvate kinase activity shows a sigmoidal dependence on PEP concentration (Fig. 4.16). ATP and certain amino acids, such as alanine and phenylalanine, compete with PEP shifting the sigmoid curve to the right and making it more sigmoidal, while F16P shifts the sigmoid curve to the left making it hyperbolic and causing a large activation at low PEP concentrations. These effects are analogous to those described in Chapter 2 for threonine deaminase and ATCase. The regulation of pyruvate kinase by F16P is another example of feedforward activation. Thus the inhibition of PFK1 and activation of F16Pase that occur in response to glucagon decrease the concentration of F16P which then suppresses the activity of pyruvate kinase. This effect is amplified by the phosphorylation of pyruvate kinase, which increases the K_m for PEP and K_a for F16P (Fig. 4.17), and decreases the K_i for ATP and alanine [57].

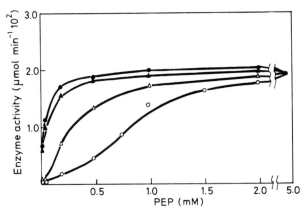

Fig. 4.16 Dependence of rat liver pyruvate kinase activity on PEP concentration. Open and filled symbols represent enzyme activity in the absence and presence of Fru-1, 6-P_2 respectively. When Fru-1, 6-P_2 was used the concentration was 5 μM. $\triangle$–$\triangle$, $\blacktriangle$–$\blacktriangle$, unphosphorylated pyruvate kinase; $\circ$–$\circ$, $\bullet$–$\bullet$, phosphorylated pyruvate kinase [57].

Alanine, phenylalanine and other amino acids are appropriate inhibitors of pyruvate kinase because they are gluconeogenic precursors. Cyclic AMP-PK also phosphorylates and activates phenylalanine hydroxylase the first enzyme in the pathway leading to the degradation of phenylalanine and tyrosine to gluconeogenic precursors [9].

4.8.3 Fatty acid synthesis and oxidation in the liver
In the fed state (low glucagon/insulin ratio) glucose is converted to pyru-vate in the cytoplasm, and then to citrate via acetyl CoA in the mitochon-dria. The citrate is transported back to the cytoplasm, where it is first

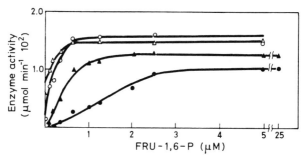

Fig. 4.17 Activity of rat liver pyruvate kinase as a function of Fru-1, 6-P_2 concentration in the presence (filled symbols) and absence (open symbols) of ATP and alanine. 0.2 mM PEP was used. When ATP and alanine were present their concentrations were 1.5 and 0.5 mM respectively. △−△, ▲−▲, unphosphorylated enzyme ○−○, ●−● phosphorylated enzyme [57].

cleaved to acetyl CoA, and then converted to long chain fatty acids. The first committed step in fatty acid synthesis is the conversion of acetyl CoA to malonyl CoA catalysed by acetyl CoA carboxylase. This enzyme has an absolute requirement for citrate [58], a further example of *feedforward activation* analogous to the regulation of glycogen synthase by G6P or pyruvate kinase by F16P.

In the fed state, fatty acid oxidation appears to be shut down because malonyl CoA levels rise, and this compound is a potent inhibitor of carnitine acyltransferase-I (CAT-I). CAT-I is located on the outer surface of the inner mitochondrial membrane, and catalyses the formation of fatty acylcarnitine esters, the first step in the transport of fatty acids to the mitochondrial matrix for oxidation [59].

However, in the fasted state (high glucagon/insulin ratio), acetyl CoA carboxylase is phosphorylated by cAMP-PK, which decreases its V_{max} and increases the K_a for citrate [9,58]. This suppresses the synthesis of malonyl CoA (and long chain fatty acids) and so relieves the inhibition on CAT-I, allowing fatty acids (mostly derived from triglyceride hydrolysis in adipose tissue − Section 4.8.1) to be oxidized to acetyl CoA.

Although the conversion of glucose to pyruvate is suppressed in the fasted state by the inhibition of PFK-I and pyruvate kinase, pyruvate is still produced in significant quantities from the breakdown of glucogenic amino acids such as alanine and serine as well as lactate. This is diverted back to PEP by two further devices. Firstly, the increased conversion of fatty acids to acetyl CoA elevates the mitochondrial acetyl CoA/CoA and NADH/NAD ratios which activate pyruvate dehydrogenase kinase. This causes phosphorylation of the pyruvate dehydrogenase complex on three serine residues, locking it into an inactive state [60,61]. Secondly, acetyl CoA is an allosteric activator of pyruvate carboxylase.

4.8.4 Control of cardiac muscle contractility by adrenalin

Although it is likely that regulation of processes such as muscle contractility, secretion, membrane permeability and transport, neurotranmission and sensory perception, growth and differentiation, and protein induction and degradation are also mediated by phosphorylation–dephosphorylation reactions, progress is still hampered by a lack of understanding of the molecular nature of the processes themselves. Nevertheless the importance of cAMP-PK has been established in several cases; for example in the control of cardiac muscle contractility by adrenalin.

Adrenalin increases both the force and rate at which the heart contracts, and the rate at which it relaxes. These effects are mediated by changes in the cytosolic concentration of Ca^{2+} and involve the phosphorylation of at least four proteins by cAMP-PK [62,63]. Phosphorylation of cardiac TN-I decreases the affinity of TN-C for Ca^{2+}, phosphorylation of *phospholamban* in the cardiac sarcoplasmic reticulum causes enhanced rates of Ca^{2+}-uptake into these vesicles, and phosphorylation of a Ca^{2+}-ATPase in the plasma membrane may increase Ca^{2+}-efflux from the cytosol to the extra-cellular space. These three reactions appear to underlie the increased rate of relaxation of heart muscle by adrenalin.

The *increase* in force and rate of contraction is brought about by the cyclic AMP dependent phosphorylation of one or more proteins in the plasma membrane. This activates the 'slow Ca^{2+} channel' 3–4 fold, and increases the rate of influx of Ca^{2+} into cardiac cells from the extracellular space [63].

4.8.6 Substrate specificity of cAMP-PK [34,64,65]

Although cAMP-PK has many physiological substrates, it is still a highly specific enzyme that phosphorylates relatively few proteins at significant rates. Even physiological substrates are only phosphorylated at one or two sites, out of many serine and threonine residues that are potentially available. The basis for its specificity has become clearer following the discovery that phosphorylation sites are usually preceded by two adjacent basic amino acids (Table 4.1), and this structural feature is now known to be critical for substrate recognition.

These findings indicate that proteins could acquire sensitivity to hormones by very few mutational events. Thus, if, by mutation, a protein acquired a pair of adjacent arginine residues N-terminal to a serine (or occasionally a threonine) located in an accessible region on the surface of the protein, that residue would be phosphorylated by cAMP-PK *in vivo*. Should that phosphorylation affect the function of the protein in a biologically useful way, then the mutation would be conserved and eventually spread through the population.

4.9 Calmodulin and cell regulation [9,18]

Neural and hormonal regulation of a variety of cells produces transient rises in the cytoplasmic concentration of Ca^{2+}. This can arise from accelerated influx of Ca^{2+} across the plasma membrane (Section 4.8.5) or from its mobilization from intracellular stores such as the mitochondrion or sarcoplasmic reticulum. Many of the biological actions of calcium ions are mediated by calmodulin (or structurally related proteins such as TN-C) in a manner that closely resembles the action of cAMP-PK (Fig. 4.14). The ability of calmodulin to function as a Ca^{2+}-dependent regulator of enzyme

Table 4.1 Amino acid sequences at the phosphorylation sites of some substrates for cyclic AMP dependent protein kinase [67]

Substrate	Source	Sequence
Phosphorylase kinase (α-subunit)	rabbit muscle	phe-arg-arg-leu-ser(P)-ile
Glycogen synthase (site 1a)	rabbit muscle	gln-trp-pro-arg-arg-ala-ser(P)-cys
Pyruvate kinase	rat liver	gly-tyr-leu-arg-arg-ala-ser(P)-val
Inhibitor-1	rabbit muscle	ile-arg-arg-arg-arg-pro-thr(P)-pro
Phosphorylase kinase (β-subunit)	rabbit muscle	arg-thr-lys-arg-ser-gly-ser(P)-val
Glycogen synthase (site 1b)	rabbit muscle	gly-ser-lys-arg-ser-asn-ser(P)-val
Phenylalanine hydroxylase	rat liver	ser-arg-lys-leu-ser(P)-asx
Troponin-I	rabbit heart	val-arg-arg-ser(P)-asp

activity is a consequence of the conformational changes that accompany the binding of Ca^{2+} by the protein, which lead to the formation of specific interaction domains. This allows calmodulin to bind to and activate many enzymes (Table 4.2).

Calmodulin activates a number of protein kinases in addition to phosphorylase kinase and the calmodulin dependent glycogen synthase kinase (Section 4.4). It activates a protein kinase that phosphorylates one of the *light* chains of myosin. This is essential for the generation of actomyosin ATPase and contraction in smooth muscle, but not in skeletal muscle. In the latter tissue, the role of *light* chain phosphorylation is less clear, although it may underlie the phenomenon of *post-twitch potentiation*, i.e. the gradual increase in strength of contraction that occurs when skeletal

Table 4.2 Some enzymes that are activated by calmodulin (9)

Enzyme	Major tissue
'High *Km*' cyclic AMP phosphodiesterase	Brain, heart
Adenylate cyclase	Brain
Calcium–magnesium ATPase	Erythrocyte and other plasma membranes
NAD Kinase	Higher plants (see Chapter 5)
Phosphorylase kinase	Muscle
Myosin light chain kinase	Muscle
Phospholamban kinase	Cardiac muscle
Tryptophan/tyrosine hydroxylase kinase	Brain
Protein phosphatase 2B	Muscle, brain

muscles are made to contract and relax repeatedly by stimulating them at low frequencies [67].

In cardiac muscle, a calmodulin dependent protein kinase phosphorylates *phospholamban* at a site distinct from that phosphorylated by cAMP-PK. Nevertheless, the calmodulin dependent phosphorylation is also associated with increased rates of Ca^{2+}-uptake into the sarcoplasm reticulum [63].

A calmodulin dependent protein kinase in nervous tissue phosphorylates both tyrosine and tryptophan hydroxylase, which regulate the synthesis of the neurotransmitters dopamine and noradrenaline, and serotonin respectively. Phosphorylation allows these enzymes to interact with an activator protein, thereby enhancing neurotransmitter synthesis [68].

Calmodulin also activates a protein phosphatase (2B) that is widely distributed in mammalian tissues, although present at particularly high concentrations in brain and skeletal muscle [36,69]. It is composed of two subunits A and B. Calmodulin binds to the catalytic A-subunit, while the B-subunit is related in structure and Ca^{2+}-binding properties to calmodulin and TN-C. The enzyme therefore resembles phosphorylase kinase (Section 4.4) in being regulated by two different but structurally related Ca^{2+}-binding proteins. The physiological role of the calmodulin dependent protein phosphatase is not clearly understood at present. In conjunction with a calmodulin dependent cyclic AMP phosphodiesterase (Table 4.2) it could represent a mechanism for deemphasizing the cyclic AMP signal in relation to the Ca^{2+}-signal. It is the most active phosphate toward the α-subunit of phosphorylase kinase and inhibitor-1 in skeletal muscle (Section 4.6) [36] but whether these proteins represent physiological substrates is unknown.

The cyclic AMP and Ca^{2+}-calmodulin systems are not only formally analogous (Fig. 4.14) but are very closely interlinked. The existence of a Ca^{2+}-calmodulin dependent adenylate cyclase and cyclic AMP phosphodiesterase in some tissues, implies that the intracellular level of cyclic AMP can be regulated by Ca^{2+}. Conversely, the cytoplasmic concentration of Ca^{2+} can sometimes be regulated by cAMP-PK, as in cardiac muscle (Section 4.8). Cyclic AMP-PK can also regulate some calmodulin dependent enzymes; for example, phosphorylase kinase and perhaps smooth muscle myosin *light* chain kinase [62,70]. Cyclic AMP-PK and Ca^{2+}-calmodulin dependent protein kinases frequently phosphorylate the same proteins, although usually at distinct sites due to the high specificity of protein kinases: examples include glycogen synthase (Fig. 4.13), phospholamban, and tyrosine hydroxylase [9].

Some hormones are capable of activating both the cyclic AMP and Ca^{2+}-calmodulin pathways. For example, adrenalin can bind to at least three types of receptor termed β, α_1 and α_2. Interaction with β and α_2 receptors leads to the activation (Section 4.6) and inhibition (Chapter 5) of adenylate cyclase respectively. Interaction with α_1 receptors is associated with an elevation of the cytoplasmic concentration of Ca^{2+} [71]. The proportions of β, α_1 and α_2 receptors in plasma membranes vary from tissue to tissue, and even within the same tissue depending on age, sex and species. In rat liver, the stimulation of glycogenolysis by adrenalin is mediated by cyclic AMP through β-receptors in immature animals, by Ca^{2+} through α_1-receptors in adult males, and by both cyclic AMP and Ca^{2+} in adult females [72]. On the other hand, the stimulation of glycogenolysis appears to be purely a β-adrenergic effect in rabbit and human liver, or in any mammalian skeletal muscle.

4.10 Synchronization of metabolic pathways by neural and hormonal stimuli

A principle that has been established for the regulation of cytoplasmic enzymes is that biodegradative enzymes are activated, whereas biosynthetic enzymes are generally inactivated by phosphorylation Table 4.3. This situation permits different metabolic pathways to be controlled by the same protein kinases and phosphatases, and the regulatory proteins that modulate their activities. Thus cyclic AMP dependent protein kinase and the Ca^{2+}-calmodulin complex (acting through several protein kinases) influence both biodegradative and biosynthetic enzymes. Similarly, the dephosphorylation events involved in the control of glycogen metabolism, glycolysis, gluconeogenesis, fatty acid synthesis, cholesterol synthesis and protein synthesis (Table 4.3) appear to be catalysed by just three enzymes, namely protein phosphatases-1, 2A and 2C [36]. However, because of their broad substrate specificities *in vitro*, the relative importance of these three enzymes in the control of different metabolic pathways *in vivo* has still to be fully evaluated in most cases.

It should be emphasized, however, that if two or more enzymes are phosphorylated and dephosphorylated by the same protein kinase and protein phosphatase, the different activities can still be regulated independently by changes in the concentrations of substrates and allosteric effectors (Section 4.8). It is therefore a device that allows for both synchronous and independent control of different enzyme activities.

4.11 The mechanism of action of insulin

Insulin is known to promote the dephosphorylation and activation of glycogen synthase [45], hydroxymethylglutaryl CoA reductase [73] and the aminoacyl-tRNA synthetase complex [74] and these effects are likely to account (at least in part) for its ability to stimulate glycogen, cholesterol and protein synthesis *in vivo*. Since these biosynthetic enzymes are regulated by protein kinases whose activities are unaffected by cyclic AMP or calmodulin (Table 4.3), this raises the question of whether such kinases mediate the actions of insulin on biosynthetic processes.

The activation of glycogen synthase by insulin in skeletal muscle appears to involve an inhibition of GSK-3 (Section 4.7.3), but the molecular mechanism is unknown. One possibility is that interaction of insulin with its receptors triggers the formation of a *mediator*, distinct from cyclic AMP or Ca^{2+}, which then inhibits GSK-3. However, evidence is accumulating that the receptors for several growth promoting hormones, such as epidermal growth factor (EGF) and insulin are actually protein kinases that become activated when they interact with hormones [75,76]. It could

Table 4.3 Activation of biodegradative enzymes and inhibition of biosynthetic enzymes by phosphorylation

	Type of protein kinase involved			
	cAMP	Ca^{2+}-calmodulin	Other	
Activation by phosphorylation				*Biodegradative pathway*
Phosphorylase	−	+	−	Glycogenolysis
Phosphorylase kinase	+	−	−	Glycogenolysis
Myosin	−	+	−	ATP hydrolysis
Triglyceride lipase	+	−	−	Lipolysis
Cholesterol esterase	+	−	−	Cholesterol ester breakdown
Phenylalanine hydroxylase	+	−	−	Aromatic amino acid breakdown
Inactivation by phosphorylation				*Biosynthetic pathway*
Glycogen synthase	+	+	+	Glycogen synthesis
Acetyl CoA carboxylase	+	−	+	Fatty acid synthesis
Pyruvate kinase (liver)	+	−	+	ATP or fatty acid synthesis
Hydroxymethyl-glutaryl CoA reductase	−	−	+	Cholesterol synthesis
Aminoacyl tRNA synthetases	−	−	+	Protein synthesis

therefore be envisaged that the receptor is a transmembrane protein in which the binding site for insulin is located on the outer surface and the protein kinase activity on the inner surface of the plasma membrane. The hormone-receptor interaction would then generate protein kinase activity directly, and might lead to the phosphorylation of intracellular proteins such as GSK-3 without the need for any intracellular mediator, akin to cyclic AMP or Ca^{2+}.

The insulin receptor and EGF receptor belong to a unique class of protein kinase that phosphorylate tyrosine rather than serine and threonine residues. This is very unusual, since phosphotyrosine represents only 0.03% of all phosphorylated amino acids in normal cells, phosphoserine and phosphothreonine accounting for the remaining 99.97% [77]. The low amounts of phosphotyrosine *in vivo* suggest that relatively few substrates for tyrosine kinases may exist. Alternatively, the physiological substrates of these enzymes may be present at extremely low concentrations.

Although the effects of insulin on processes such as glucose transport and glycogen synthesis in skeletal muscle, or fatty acid synthesis in adipose tissue, are clearly independent of cyclic AMP, insulin is also capable of *lowering* cyclic AMP levels *in vivo*, provided they have previously been raised by cyclic AMP elevating hormones [29]. This is an important metabolic action of insulin, enabling it to antagonize the various actions of glucagon in mammalian liver and the action of adrenalin on triglyceride lipase in adipose tissue (Section 4.8). However, whether insulin inhibits adenylate cyclase or activates a cyclic AMP phosphodiesterase, or both, is still uncertain, and there is not yet general agreement as to the molecular mechanism(s) involved.

Insulin also causes *increased* phosphorylation of several proteins (on serine residues), indicating that at least one protein kinase must be activated or at least one protein phosphatase inhibited by this hormone. Whether increased and decreased phosphorylation of proteins by insulin can be explained by a common mechanism remains to be determined. The molecular action of insulin is the major outstanding problem in the hormonal control of cellular metabolism. For further discussion and an alternative view of this controversial topic see [78].

4.12 Summary

The cyclic interconversion of key enzymes between phosphorylated and dephosphorylated forms is an extremely versatile mechanism for reversibly altering their activities, and in mammalian cells may be almost as common as allosteric regulation. Because the phosphorylation and dephosphorylation reactions are catalysed by separate *converter* enzymes, they each involve the action of one enzyme upon another and are therefore defined as *cascade* systems [79]. In *monocyclic cascades* allosteric effectors can bind to the *interconvertible* enzyme or to either or both of the converter enzymes. Through these interactions, a change in the concentration of one or more effectors will lead automatically to a change in the relative activities of the converter enzymes, and hence to an adjustment in the steady state level of phosphorylation and activity of the *interconvertible enzyme*. Thus monocyclic cascades act as metabolic integration systems, which can sense changes in the concentrations of many different metabolites. However, theoretical analyses have shown that monocyclic cascade systems have several further properties of potential importance in cellular regulation [79].

(a) They have an enormous capacity for *signal amplification*, that allow large changes in the phosphorylation state of the interconvertible enzyme to occur in response to very small alterations in the concentrations of allosteric modifiers.

(b) They can function as *rate amplifiers*, and respond to changes in allosteric effector concentrations within milliseconds.

(c) They can exhibit great flexibility of response, depending on whether an allosteric modifier activates or inhibits a converter enzyme, whether the converter enzymes are activated or inhibited by the same or different effectors, or whether the effectors bind to the interconvertible enzyme thereby altering the rate at which it is phosphorylated and/or dephosphorylated. In situations where an allosteric modifier activates one converter enzyme and inhibits another (e.g. activation of cAMP-PK by cyclic AMP and inhibition of protein phosphatase-1 by cyclic AMP through the phosphorylation of inhibitor-1 by cAMP-PK (Fig. 4.12) the capacity for signal amplification is greatly increased, and there is a *sigmoidal* relationship between the steady state level of *interconvertible enzyme* phosphorylation and effector concentration. The finding that interconvertible enzymes are frequently phosphorylated at several sites by the same protein kinases or at multiple sites by several different protein kinases (e.g. glycogen synthase) adds another dimension to the cascade form of control, since these *multi-site phosphorylations* can have synergistic effects on the kinetic parameters, and also affect the regulation of the dephosphorylation process.

When the active form of an *interconvertible enzyme* in one cascade serves as a converter enzyme in a second phosphorylation–dephosphorylation cycle, the two cycles become coupled to form a *bicyclic cascade*. This increases exponentially the flexibility of response and the capacity for *signal* and *rate amplification*.

Mono, bi, and multicyclic cascades may also serve two further functions. Firstly they sometimes link in sequence two or more enzymes whose intracellular concentrations differ enormously, increasing the amplification potential of the system and allowing the ultimate *interconvertible enzyme* in the sequence to be regulated by *metabolites* that vary greatly in concentration. For example, the molar concentrations of cAMP-PK, phosphorylase kinase and phosphorylase are $0.2 \ \mu M$, $2.5 \ \mu M$ and $80 \ \mu M$ respectively in white 'fast twitch' muscle fibres [34]. It would therefore be impossible for phosphorylase to be activated directly by cyclic AMP whose intracellular level rarely exceeds $2–3 \ \mu M$. Secondly, the surface area of any one protein is not unlimited, and it may not be feasible to incorporate into a single molecule all the controls that are required of an enzyme that occupies an important position in metabolism.

However, it is now evident that the reversible phosphorylation of enzymes is the basis of a complex network of interlocking cascades that can respond to common biological signals (hormones, nerve impulses, etc.), allowing coordinated and synchronized control of many biochemical functions.

References

[1] Krebs, E. G. and Fischer, E. H. (1956), *Biochim. et Biophys. Acta*, **20**, 150−157.

[2] Krebs, E. G., Graves, D. J. and Fischer, E. H. (1959), *J. Biol. Chem.*, **234**, 2867.

[3] Friedman, D. L. and Larner, J. (1963), *Biochemistry*, 2, 669−675.

[4] Walsh, D. A., Perkins, J. P. and Krebs, E. G. (1968), *J. Biol. Chem.*, **243**, 3763−3765.

[5] Karnieli, E., Zarnowski, M. J., Hissin, P. J., Simpson, I. A., Salans, L. B. and Cushman, S. W. (1981), *J. Biol. Chem.*, **256**, 4772−4777.

[6] Cori, G. T., Colowick, S. P. and Cori, C. F. (1938), *J. Biol. Chem.*, **123**, 381−389.

[7] Cori, G. T. and Green, A. A. (1945), *J. Biol. Chem.*, **151**, 31−38.

[8] Titani, K., Cohen, P., Walsh, K. A. and Neurath, H. (1975) *FEBS Lett.*, **55**, 120−123.

[9] Cohen, P. (1982), *Nature*, **296**, 613−620.

[10] Bagshaw, C. R. (1982), *Muscle Contraction*, Chapman and Hall, London.

[11] Danforth, W. H. and Helmreich, E. (1964), *J. Biol. Chem.*, **239**, 3133−3138.

[12] Danforth, W. H. and Lyon, J. B. (1964), *J. Biol. Chem.*, **239**, 4047−4050.

[13] Helmreich, E. and Cori, C. F. (1966), *Adv. Enzyme Reg.*, **3**, 91−107.

[14] Cohen, P. (1980), *Eur. J. Biochem.*, **111**, 563−574.

[15] Cohen, P., Klee, C. B., Picton, C. and Shenolikar, S. (1980), *Ann. N.Y. Acad. Sci.*, **356**, 151−161.

[16] Skuster, J. F., Jesse-Chan, K. F. and Graves, D. (1980), *J. Biol. Chem.*, **255**, 2203−2210.

[17] Cohen, P., Burchell, A., Foulkes, J. G., Cohen, P. T. W., Vanaman, T. C. and Nairn, A. C. (1978), *FEBS Lett.*, **92**, 287−293.

[18] Klee, C. B., Crouch, T. H. and Richman, P. G. (1980), *Ann. Rev. Biochem.*, **49**, 489−515.

[19] Kilhofer, M. C., Gerard, D., and Demaille, J. G. (1980), *FEBS Lett.*, **120**, 99−103.

[20] Picton, C., Klee, C. B. and Cohen, P. (1980), *Eur. J. Biochem.*, **111**, 553−561.

[21] Helmreich, E. and Cori, C. F. (1964), *J. Biol. Chem.*, **239**, 2440−2445.

[22] Lowry, O. H., Schultz, D. W. and Passoneau, J. V. (1964), *J. Biol. Chem.*, **239**, 1947−1953.

[23] Denton, R. M. and Pogson, C. (1976), *Metabolic Regulation*, Chapman and Hall, London.

[24] Gadian, D. G. and Radda, G. K. (1981), *Ann. Rev. Biochem.*, **50**, 69−83.

[25] Schulman, R. G. (1982), *Scientific American*, **248**, 76−83.

[26] Danforth, W. H., Helmreich, E. and Cori, C. F. (1962), *Proc. Nat. Acad. Sci* (USA), **48**, 1191−1199.

[27] Karpatkin, S., Helmreich, E. and Cori, C. F. (1964), *J. Biol. Chem.*, **239**, 3139–3145.
[28] Gratecos, D., Detwiler, T. and Fischer, E. H. (1974), in *Metabolic Interconversions of Enzymes* (1973), (eds E. H. Fischer, E. G. Krebs, H. Neurath and E. R. Stadtman), Springer-Verlag, Heidelberg, pp. 43–52.
[29] Robison, G. A., Butcher, R. W. and Sutherland, E. W. (1971), *Cyclic AMP*, Academic Press, New York and London.
[30] Ross, E. M. and Gilman, A. G. (1980), *Ann. Rev. Biochem.*, **49**, 533–564.
[31] Kahn, R. A., Katada, T., Bokoch, G. M., Northup, J. K. and Gilman A. G. (1983), in *Post-translational Covalent Modifications of Proteins for Function*, (ed. B. C. Johnson), Academic Press, New York and London, in press.
[32] Pedersen, S. E. and Ross, E. M. (1982), *Proc. Nat. Acad. Sci.*, USA, **79**, 7228–7232.
[33] Krebs, E. G. and Beavo, J. A. (1979), *Ann. Rev. Biochem.*, **48**, 923–959.
[34] Cohen, P. (1978), *Curr. Top. Cell Reg.*, **14**, 117–196.
[35] Singh, T. J., Akatsuka, A. and Huang, K. P. (1982), *J. Biol. Chem.*, **257**, 13379–13384.
[36] Ingebritsen, T. S. and Cohen, P. (1983), *Science*, in press.
[37] Aitken, A., Bilham, T. and Cohen, P. (1982), *Eur. J. Biochem.*, **126**, 235–246.
[38] Foulkes, J. G. and Cohen, P. (1979), *Eur. J. Biochem.*, **97**, 251–256.
[39] Cohen, P., Yellowlees, D., Aitken, A., Donella-Deana, A., Hemmings, B. A. and Parker, P. J. (1982), *Eur. J. Biochem.*, **124**, 21–35.
[40] Picton, C., Aitken, A., Bilham, T. and Cohen, P. (1982), *Eur. J. Biochem.*, **124**, 37–45.
[41] Parker, P. J., Embi, N., Caudwell, F. B. and Cohen, P. (1982), *Eur. J. Biochem.*, **124**, 47–55.
[42] Woodgett, J. R., Tonks, N. K. and Cohen, P. (1982), *FEBS Lett.*, **148**, 5–11.
[43] Roach, P. J. (1982), *Curr. Top. Cell. Reg.*, **20**, 45–105.
[44] Roach, P. J., DePaoli-Roach, A. A. and Larner, J. (1978), *J. Cyc. Nuc. Res.*, **4**, 245–257.
[45] Parker, P. J., Caudwell, F. B. and Cohen, P. (1983), *Eur. J. Biochem.*, **130**, 227–234.
[46] Villar-Palasi, C. and Larner, J. (1961), *Biochim. et Biophys. Acta*, **139**, 171–173.
[47] Danforth, W. H. (1965), *J. Biol. Chem.*, **240**, 588–593.
[48] Picton, C., Woodgett, J. R., Hemmings, B. A. and Cohen, P. (1982), *FEBS Lett.*, **150**, 191–196.
[49] Kuo, J. F. and Greengard, P. (1969), *Proc. Nat. Acad. Sic.* (USA), **64**, 1349–1355.
[50] Krebs, E. G. (1972), *Curr. Top. Cell. Reg.*, **5**, 99–133.
[51] Fredrickson, G., Stralfors, P., Nilsson, N. O. and Belfrage, P. (1981), *J. Biol. Chem.*, **256**, 6311–6320.
[52] Boyd, G. S. and Gorban, A. N. S. (1980), in *Molecular Aspects of Cellular Regulation*, (ed. P. Cohen), Vol. 1, Elsevier Biomedical, Amsterdam, pp. 95–134.
[53] Cook, K. G., Yeaman, S. J., Stralfors, P., Fredrickson, G. and Belfrage, P. (1982), *Eur. J. Biochem.*, **125**, 245–249.

[54] Hers, H. G., and Van-Schaftingen, E. (1982), *Biochem. J.*, **206**, 1–12.

[55] El-Maghrabi, M. R., Claus, T. H., Pilkis, J., Fox, E. and Pilkis, S. J. (1982), *J. Biol. Chem.*, **257**, 7603–7607.

[56] LaPorte, D. C. and Koshland, D. E. (1982), *Nature*, **300**, 458–460.

[57] Engstrom, L. (1980), in *Molecular Aspects of Cellular Regulation*, (ed. P. Cohen), Vol. 1, Elsevier Biomedical, Amsterdam, pp. 11–31.

[58] Hardie, D. G. (1980), in *Molecular Aspects of Cellular Regulation*, (ed. P. Cohen), Vol. 1, Elsevier Biomedical, Amsterdam, pp. 33–62.

[59] McGarry, J. D. and Foster, D. W. (1980), *Ann. Rev. Biochem.*, **49**, 395–420.

[60] Denton, R. N. and Hughes, W. A. (1978), *Int. J. Biochem.*, **9**, 545–552.

[61] Randle, P. J. (1983), *Phil-Trans. Roy. Soc. Lond.* Series B, (in press).

[62] England, P. J. (1980), in *Molecular Aspects of Cellular Regulation*, (ed. P. Cohen), Vol. 1, Elsevier Biomedical, Amsterdam, pp. 153–173.

[63] Haiech, J. and Demaille, J. G. (1983), *Phil. Trans. Roy. Soc. Lond.* Series B, (in press).

[64] Cohen, P. (1980), in *Molecular Aspects of Cellular Regulation*, (ed. P. Cohen), Vol. 1, Elsevier Biomedical, Amsterdam, pp. 255–268.

[65] Nimmo, H. G., and Cohen, P. (1977), *Adv. Cyc. Nuc. Res.*, **8**, 145–266.

[66] Cohen, P., Aitken, A., Damuni, Z. *et al.* (1983), in *Post-translational Covalent Modifications of Proteins for Function*, (ed. B. C. Johnson), Academic Press, New York and London, *in press*.

[67] Klug, G. A., Botterman, B. R. and Stull, J. T. (1982), *J. Biol. Chem.*, **257**, 4688–4690.

[68] Yamauchi, T., Nakata, H. and Fujisawa, H. (1981), *J. Biol. Chem.*, **256**, 5404–5409.

[69] Stewart, A. A., Ingebritsen, T. S., Manalan, A., Klee, C. B. and Cohen, P. (1982), *FEBS Lett.*, **137**, 80–84.

[70] Adelstein, R. S. and Eisenberg, E. A. (1980), *Ann. Rev. Biochem.*, **49**, 921–956.

[71] Exton, J. H. (1979), *J. Cyc. Nuc. Res.*, **5**, 277–287.

[72] Studer, R. K. and Borle, A. B. (1982), *J. Biol. Chem.*, **257**, 7987–7993.

[73] Ingebritsen, T. S. and Gibson, D. M., (1980), in *Molecular Aspects of Cellular Regulation*, (ed. P. Cohen), Vol. 1, Elsevier Biomedical, Amsterdam, pp. 63–93.

[74] Damuni, Z. and Cohen, P. (1982), *Eur. J. Biochem.*, **129**, 57–65.

[75] Buhrow, S. A., Cohen, S. and Staros, J. V. (1982), *J. Biol. Chem.*, **257**, 4019–4022.

[76] Kasuga, M., Zick, Y., Blithe, D. L., Crettaz, M. and Kahn, C. R. (1982), *Nature*, **298**, 667–669.

[77] Sefton, B. M., Hunter, T., Beemen, K. and Eckhart, W. (1980), *Cell*, **20**, 807–816.

[78] Denton, R. M., Brownsey, R. W. and Belsham, G. J. (1981), *Diabetologia*, **21**, 347–362.

[79] Chock, P. B., Rhee, S. G. and Stadtman, E. R. (1980), *Ann. Rev. Biochem.*, **49**, 813–843.

5 Control of enzyme activity by covalent modifications other than limited proteolysis or phosphorylation

Although only 20 amino acids are specified in the genetic code, it is now evident that post-translational modification of proteins is very substantial. Over 150 amino acid derivatives have been recognized *in vivo* [1], of which phosphorylation, glycosylation and acetylation of α-amino groups are the most frequent. This chapter reviews the systems in which covalent modifications, other than limited proteolysis and phosphorylation, have been shown to be involved in the control of enzyme activity.

5.1 The regulation of *E. coli* glutamine synthetase activity by adenylylation and uridylylation

5.1.1 Feedback inhibition
Glutamine is of central importance in the nitrogen metabolism of micro-organisms, since its amide group, and not ammonia, is the preferred nitrogen donor in the synthesis of tryptophan, AMP, CTP, glucosamine-6-phosphate, histidine and carbamyl phosphate. In addition, the α-amino group can serve as a source of nitrogen in the synthesis of glycine and alanine, through the action of specific transaminases. As the first enzyme in a highly branched pathway, leading to the synthesis of a wide range of diverse metabolites, glutamine synthetase should be a prime target for metabolic control. However, it is the extraordinary complexity of the mechanism, solved by Earl Stadtman and his coworkers, that has been totally unexpected.

E. coli possess just a single glutamine synthetase, which was shown to be inhibited *in vitro* by all eight end products of glutamine metabolism (Fig. 5.1), although 50–60 other metabolites tested were without effect [2]. While only partial inhibition could be achieved by any one end product, even at saturating concentrations, the effects of different inhibitors were additive, so that almost total inhibition was observed when all eight end products were present in excess. This phenomenon, referred to as *cumulative feedback inhibition*, allowed three inferences to be made. Each inhibitor bound to a separate site on the enzyme, all the inhibitors could bind simultaneously, and the binding of any one inhibitor did not modify the binding of any other inhibitor by the enzyme. These ideas have been confirmed by direct binding studies in the case of AMP and tryptophan [4], which showed that glutamine synthetase could bind 12 molecules of each feedback regulator. This is consistent with the structure of the enzyme, which is formed from 12 identical M_r 50 000 subunits, arranged as two hexagonal layers, stacked one upon the other (Fig. 5.2). In view of the relatively modest size of the subunit, the number of different binding sites that must exist is rather remarkable.

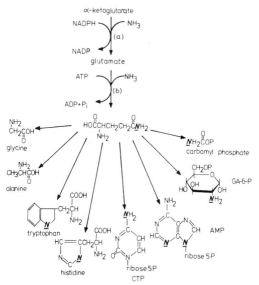

Fig. 5.1 Metabolic fates of glutamine. The nitrogen atoms donated by glutamine are in bold italics underlined [3]. (a) glutamate dehydrogenase, (b) glutamine synthetase.

By analogy with aspartokinase (Chapter 2), cumulative inhibitions, if they are to be effective, must of course be supplemented with secondary controls on the first step of each *branch* of the pathway, by the ultimate end product of that branch. These have been established in most cases, and discussed for tryptophan and CTP in Chapter 2.

5.1.2 Adenylylation of glutamine synthetase

In the mid 1960s, observations from two laboratories began to show that a further mechanism for regulating glutamine synthetase activity must exist. In Germany, Helmut Holzer [6] showed that if *E. coli* were grown on a nitrogen source other than NH_4^+ the cells maintained high levels of glutamine synthetase activity. If, however, NH_4^+ salts were added to the growth medium, a ten-fold reduction in activity occurred within a minute, which could not be due to repression of the synthesis of the enzyme, as the generation time of the cells was 50 min. NH_4^+ must therefore be promoting an inactivation of pre-existing glutamine synthetase molecules. The effect was specific for NH_4^+, but NH_4^+ itself was a substrate and not an inhibitor *in vitro*. A search for the factor responsible, led to the discovery of an enzyme that catalysed the inactivation of glutamine synthetase *in vitro* if supplemented with ATP, Mg^{2+} and glutamine. It was a reasonable hypothesis, that NH_4^+ promoted an increase in the intracellular concentration of glutamine, at the expense of α-ketoglutarate and glutamate (Fig. 5.1) which then activated the enzyme, that inactivated glutamine synthetase.

At the same time, Earl Stadtman in the USA had discovered that glutamine synthetases isolated from *E. coli* grown under different conditions, varied not only in their specific activity, but in their Me^{2+} specificity, and

73

Fig. 5.2 Arrangement of subunits in *E. coli* glutamine synthetase [3,5].

response to feedback inhibitors [3,4,7]. He then showed that the different forms had varying quantities of covalently bound adenylic acid (AMP). It became clear, that the two groups were studying the same effect, and that Holzer's inactivating protein was an adenylyl transferase (termed ATase) which catalysed the covalent attachment of AMP to a unique tyrosyl group on each of the 12 subunits of glutamine synthetase (Fig. 5.3) [8,9,10]:

$$(E_0) + 12ATP \xrightarrow{Mg^{2+}} (E_{12}) + 12PP_i \qquad (5.1)$$

deadenylylated   adenylylated
glutamine   glutamine
synthetase   synthetase

Conversion of fully deadenylylated (E_0) to fully adenylylated (E_{12}) enzyme, transformed glutamine synthetase from an active, Mg^{2+} dependent form, having a pH optimum of 7.6, to a less active, Mn^{2+} dependent form, optimally active at pH 6.5 [4]. The Mg^{2+} dependent E_0 enzyme was sensitive to inhibition by alanine, glycine and AMP, whereas the Mn^{2+} dependent E_{12} enzyme was most sensitive to inhibition by histidine, tryptophan and CTP [7].

Mg^{2+} and Mn^{2+} presumably function in glutamine synthetase, as in other ATP requiring enzymes as Mg^{2+}–ATP and Mn^{2+}–ATP complexes respectively, as evidenced by the requirement for equal concentrations of Me^{2+} and ATP to achieve optimal activity. Since *E. coli* grow perfectly well on synthetic media without the addition of Mn^{2+}, and intracellular Mn^{2+} concentrations are only in the μM range, whereas Mg^{2+} and ATP are in the mM range, it seems probable that only the Mg^{2+} linked activity is physiologically significant. The conversion of E_0 to E_{12} should therefore probably be regarded as the transformation of an active form sensitive to inhibition by alanine, glycine and AMP, to an inactive enzyme.

However, the conversion of E_0 to E_{12} would only take place under

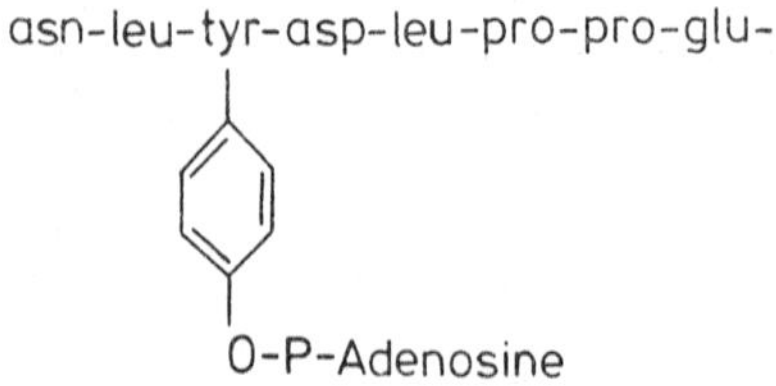

asn-leu-tyr-asp-leu-pro-pro-glu-

O-P-Adenosine

Fig. 5.3 Amino acid sequence at the site of adenylylation of glutamine synthetase [11].

extreme conditions; i.e. when cells grown under NH_4^+ starvation come in contact with media rich in NH_4^+. Glutamine synthetase will commonly be found at intermediate states of adenylylation. This raises the question of the possible existence and role of *hybrid* molecules, in which only some of the 12 subunits are adenylylated.

The number of possible forms of glutamine synthetase bearing in mind not only the number of adenylyl groups per molecule, but also their position relative to one another, within the two hexagonal rings, is 382! That hybrid molecules with distinctive properties do exist is shown in Fig. 5.4. The specific activity of glutamine synthetase is *not* a linear function of the state of adenylylation, and sharper changes in activity occur as the enzymes tend towards either E_0 or E_{12}. Other kinetic parameters of partially adenylylated preparations, and also their resistance to denaturation, differ from equivalent *mixtures* of E_0 and E_{12} enzymes [12,13]. The activities of any subunit are clearly influenced by the state of adenylylation of neighbouring subunits, presumably through allosteric transitions transmitted across the subunit contacts, analogous to those proposed for ATCase (Chapter 2). The original preparations of glutamine synthetase which showed cumulative feedback inhibition to all eight feedback inhibitors, and were more sensitive to inhibition by low concentrations of these regulators than E_0 or E_{12} were in all probability partially adenylylated species [2].

5.1.3 Control of adenylylation and deadenylylation by reversible uridylylation

When *E. coli* were transferred from an NH_4^+ rich to an NH_4^+ poor medium, glutamine synthetase reactivated rapidly, showing that adenylylation was reversible [6]. This focused attention on the control of deadenylylation, sparking off the remarkable series of findings, which are listed below.

(a) Deadenylylation proceeds through a phosphorylytic cleavage of the tyrosyl-O-AMP linkage [14]

$$E_{12}(AMP) + 12P_i \longrightarrow E_0 + 12ADP \qquad (5.2)$$

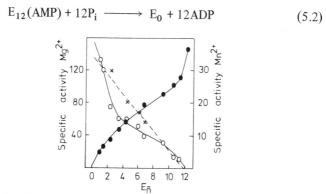

Fig. 5.4 Relationship between the state of adenylylation and the specific activity of glutamine synthetase. (o—o)–Mg^{2+} linked activity; (•—•)–Mn^{2+} linked activity; (X———X)–Mg^{2+} linked activity obtained by mixing E_0 and E_{12} in varying proportions [12].

(b) Deadenylylation requires two distinct proteins termed P_I and P_{II} [15].

(c) The P_I protein is identical with ATase. The same enzyme therefore catalyses both adenylylation and deadenylylation [16].

(d) ATase ($M_r \sim 130\,000$) is a single *bifunctional* polypeptide with separate catalytic sites for the adenylylation and deadenylylation reactions [17].

(e) The P_{II} protein exists in two forms termed $P_{II}-A$ and $P_{II}-D$. For the adenylylation reaction glutamine synthetase and ATP are substrates and $P_{II}-A$ and glutamine are activators. $P_{II}-A$ activates the reaction by increasing the affinities of glutamine synthetase (16-fold) and glutamine (10-fold) for ATase. $P_{II}-A$ has no effect on the V_{max} of ATase. However glutamine increases the V_{max} (20-fold) and the affinities for glutamine synthetase (6-fold) and $P_{II}-A$ (10-fold) [3]. Deadenylylation has an absolute requirement for $P_{II}-D$, as well as ATP and α-ketoglutarate [18].

(f) The conversion of $P_{II}-A$ to $P_{II}-D$ requires UTP, a further enzyme, and either ATP or α-ketoglutarate. The presence of both ATP and α-ketoglutarate stimulates the reaction a further 2–3 fold. The formation of $P_{II}-D$ is inhibited by glutamine and also P_i [19,20].

(g) The P_{II} protein, M_r 44 000, is formed from four identical subunits. The conversion of $P_{II}-A$ to $P_{II}-D$ involves the covalent attachment of UMP to one of the two tyrosyl residues on each subunit, catalysed by a specific uridylyl transferase (UTase) [21].

$$P_{II}-A + 4UTP \xrightarrow{\text{UTase, Mg}^{2+}} P_{II}-D(UMP) + 4PP_i \quad (5.3)$$

(h) Deuridylylation is catalysed by a uridylyl removing enzyme (UR-enzyme). This enzyme can act in the presence of Mn^{2+} alone, or with Mg^{2+} in the presence of either α-ketoglutarate or ATP. The presence of both ATP and α-ketoglutarate stimulates the Mg^{2+} linked UR-enzyme activity a further 2-fold [22]

$$P_{II}-D(UMP) + 4H_2O \xrightarrow{\text{UR-enzyme, Me}^{2+}} P_{II}-A + 4UMP \quad (5.4)$$

(i) A single polypeptide ($M_r \sim 98\,000$) contains both UTase and UR-enzymes. This bifunctionality is analogous to that exhibited by ATase [23].

The net result of all these discoveries is that glutamine synthetase is regulated by the two oppositely directed *cascade* systems depicted in Fig. 5.5.

Inactivation of glutamine synthetase (adenylylation) is initiated by UR-enzyme, which catalyses the conversion of $P_{II}-D$ to $P_{II}-A$ (deuridylylation). $P_{II}-A$ interacts with ATase, stimulating its capacity to catalyse the adenylylation of glutamine synthetase. Similarly, activation of glutamine synthetase (deadenylylation) is initiated by UTase, which converts $P_{II}-A$ to $P_{II}-D$ (uridylylation). Interaction of $P_{II}-D$ with ATase stimulates the capacity of ATase to catalyse deadenylylation. The allosteric transitions initiated by the P_{II}-protein therefore represent a remarkable

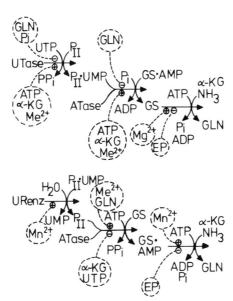

Fig. 5.5 Cascade control of glutamine synthetase (GS). A—activation (deadenylation) of GS: B—inactivation (adenylation) of GS; EP, end products; KG—α-ketoglutarate; GLN—glutamine [21,22].

situation in which the directionality as well as the rate of an enzyme catalysed reaction is affected.

The steady state level of adenylylation is under the control of the metabolites α-ketoglutarate, glutamine, UTP, ATP and P_i. Other reports, that RNA fractions can modify P_{II} activity [15], while glycolytic pathway intermediates inhibit ATase [24], suggests that the list of physiological regulators may be by no means complete. The opposing effects of α-ketoglutarate, glutamine and UTP on the two *cascades* ensures that the state of adenylylation will be very sensitive to small changes in the relative concentrations of these effectors. It also ensures that the activities of the various modifying enzymes are closely controlled. This minimizes haphazard coupling of reactions 5.1 and 5.2, and 5.3 and 5.4, which if uncontrolled would lead to wasteful energy loss through the hydrolysis of ATP and UTP. Reconstitution experiments using the isolated protein components and some effectors, have demonstrated that the regulation of glutamine synthetase by adenylylation and uridylylation consumes less than 1% of the ATP that is utilized directly by glutamine synthetase in the synthesis of glutamine (Fig. 5.1). Enzyme regulation by reversible covalent modification, of course, uses much less energy than would be required for the synthesis of a new enzyme molecule [25].

While the state of adenylylation is regulated by α-ketoglutarate, glutamine, UTP, ATP and P_i, the activity of glutamine synthetase is regulated by eight different effectors, namely the end products of glutamine metabolism (Fig. 5.1). The glutamine/α-ketoglutarate ratio may be thought of as a sensitive index of the NH_4^+ concentration [26]. When NH_4^+ is in excess, glutamine will be formed at the expense of α-ketoglutarate (Fig. 5.1). The increased glutamine/α-ketoglutarate ratio may then result in complete adenylylation of glutamine synthetase. The important role of NH_4^+ may be explained not only by the increased formation of glutamine under these conditions, but by the finding that NH_4^+ at high concentrations can replace the amide groups of glutamine as the nitrogen donor in the biosynthetic reactions shown in Fig. 5.1. Conversely, under nitrogen starvation, the glutamine/α-ketoglutarate ratio will decrease and complete deadenylylation may take place. This should allow any NH_4^+ that becomes available to be scavenged effectively. Under most growth conditions, intermediate states of adenylylation, determined by the relative levels of all effectors, will predominate. It seems likely that a whole range of hybrid molecules can form *in vivo* which show different patterns of sensitivity to the eight feedback inhibitors.

The net effect of UTP is to stimulate deadenylylation and increase glutamine synthetase activity, and therefore apparently opposes the action of the feedback inhibitor CTP. Perhaps this is a further example of the need to balance the synthesis of different nucleoside triphosphates (Chapter 2) or perhaps it reflects the need of the cell to generate a high nucleic acid synthesizing capacity.

Adenylylation can also be thought of as an indirect means of achieving a glutamine dependent inactivation of glutamine synthetase. The potential advantages of covalent modification as opposed to a direct allosteric effect were discussed fully in Chapter 4, but it is nevertheless curious that only Gram-negative bacteria and one Gram-positive bacterium, *Streptomyces cattleya* [27] have evolved this highly sophisticated type of regulation. In *Bacillus subtilis*, glutamine itself inhibits glutamine synthetase directly and there appears to be no control by covalent modification [28]. The regulation of the *E. coli* enzyme nevertheless demonstrates that control mechanisms exist in *prokaryotes* which compare in complexity to those found in *eukaryotes*. No examples of enzyme regulation by adenylylation or uridylylation have so far been found in eukaryotic cells.

5.2 The regulation of citrate lyase in anaerobic bacteria by acetylation–deacetylation [29]

Citrate is abundant in Nature and is found in many plants, in milk and in bones. It is therefore not surprising that many bacterial species are capable of using it for growth. However, a special mechanism is required to degrade it in anaerobic bacteria, because they do not usually contain a complete tricarboxylic acid cycle. This is accomplished by the enzymes citrate lyase and oxaloacetate decarboxylase which catalyse the following two reactions:

$$citrate \rightleftharpoons oxaloacetate + acetate$$

$$oxaloacetate \longrightarrow pyruvate + CO_2$$

Citrate lyase contains acetyl residues, linked by a thioester bond to a derivative of coenzyme A, which participate directly in the enzymic reaction. Citrate is first activated in a transferase reaction and then cleaved in such a way that the acetylated form of the enzyme is regenerated:

$$\text{enzyme-S-acetyl} + \text{citrate} \longrightarrow \text{enzyme-S-citroyl} + \text{acetate}$$

$$\text{enzyme-S-citroyl} \longrightarrow \text{enzyme-S-acetyl} + \text{oxaloacetate}$$

Citrate lyase is composed of three types of subunit, M_r 54 000, 32 000 and 10 000, and the molecule contains six copies of each component. The largest subunit catalyses the transferase reaction, the M_r 32 000 species is involved in the cleavage reaction, while the smallest subunit contains the acetyl group.

In microorganisms where citrate lyase is the only enzyme that utilizes citrate, there is no need for its activity to be controlled. For example in *Streptococcus diacetilactis*, which requires L-glutamate for growth, the enzymes that form glutamate from citrate are absent, and citrate lyase activity is not regulated. However, many anaerobes are able to synthesize glutamate, and contain citrate synthase the antagonistic enzyme to citrate lyase. In this situation the activity of citrate lyase must therefore be closely controlled, otherwise it would interfere with the synthesis of glutamate and glutamine under conditions of low citrate availability. An organism of this type is *Rhodopseudomonas gelatinosa*, which decomposes citrate very rapidly under anaerobic conditions in the light. It has been shown that citrate lyase is rapidly converted from the inactive SH form to the active S-acetyl form as soon as citrate is available, and deacetylated when citrate is exhausted (Fig. 5.6).

Acetylation is carried out by an enzyme called citrate lyase ligase and deacetylation by citrate lyase deacetylase. The ligase (M_r 41 000) exists in an inactive form in the absence of citrate, but is rapidly converted to an active form when citrate is added to the culture medium. The molecular mechanism for activation is unknown. The deacetylase is strongly inhibited by L-glutamate (Fig. 5.6), ensuring that citrate lyase remains active, if there is sufficient glutamate for glutamine synthesis.

No enzyme in mammalian cells is known to be regulated by reversible acetylation. However, the reversible acetylation of histones in the nucleus

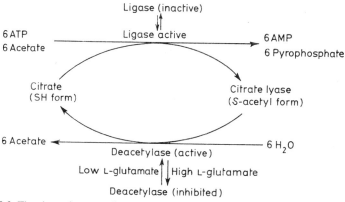

Fig. 5.6 The citrate lyase regulatory system of *Rhodopseudomonas gelatinosa* [29].

is well documented, and may play a role in the regulation of gene transcription. Acetylation of histones occurs on lysine residues, and not on sulphydryl groups [30].

5.3 Light activation of chloroplast enzymes by thiol–disulphide interchange [31]

During the day, plants obtain energy from absorbed light and use it to convert CO_2 to energy reserves, such as starch, using a pathway known as the *reductive* pentose phosphate cycle (Fig. 5.7). At night however, the plant becomes, in biochemical terms, an 'animal', and mobilizes the carbohydrate it synthesized during the day by means of the glycolytic and *oxidative* pentose phosphate cycles.

Since the 'light' (photosynthetic) and 'dark' (degradative) metabolic pathways are both localized within chloroplasts, it is clearly essential that mechanisms should exist for controlling these opposing reactions. There is now good evidence that at least four enzymes of the *reductive* pentose phosphate cycle [NADPH linked glyceraldehyde 3 phosphate dehydrogenase (G3PD), sedoheptulose 1:7 bisphosphatase (S17Pase), fructose 1:6 bisphosphatase (F16Pase) and phosphoribulokinase (PRK)] undergo a rapid activation when chloroplasts are exposed to light (Fig. 5.8). Conversely, phosphofructokinase (PFK) in the glycolytic pathway and NADP linked glucose-6-phosphate dehydrogenase (G6PD) in the *oxidative* pentose phosphate pathway are inactivated by light.

The mechanism by which light may regulate these enzymes is illustrated in Fig. 5.8. In the light, electrons from chlorophyll are used to reduce the iron atom at the active site of a protein termed ferredoxin, and the reoxidation of this iron atom is coupled to the reduction of a further protein termed thioredoxin, by the enzyme ferredoxin–thioredoxin reductase (Fig. 5.8). Thioredoxin (M_r 12 000) contains a redox-active disulphide bridge, which in its reduced SH form can function as a protein-disulphide

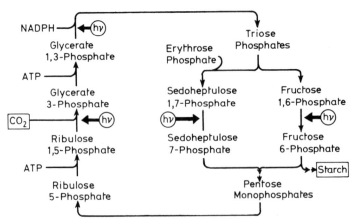

Fig. 5.7 Role of light (hν) in the activation of enzymes of the reductive pentose phosphate cycle [31].

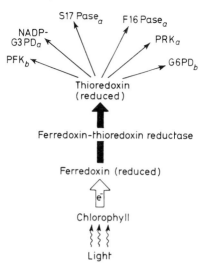

Fig. 5.8 Role of thioredoxin in the regulation of chloroplast enzymes [32]; a, activated; b, inactivated. Enzyme abbreviations are defined in the text.

reductase. It is therefore assumed that activation of the enzymes of the *reductive* pentose phosphate cycle results from the conversion of inactive $-S-S-$ to active SH enzymes. This idea is supported by the finding that thiols such as dithiothreitol (DTT) can mimic the action of thioredoxin. Conversely, the inactivation of PFK and G6PD may occur as a result of their conversion from active $-S-S-$ to inactive SH enzymes, since the effects of light can be mimicked by DTT.

Enzymes reduced by thioredoxin in the light are reoxidized to $-S-S-$ forms in the dark. This may be catalysed by compounds such as oxidized glutathione, but the mechanisms involved are not yet fully elucidated.

Regulation by thiol−disulphide interchange in plants, like phosphorylation−dephosphorylation in mammals, is supplemented by a variety of allosteric controls. For example, Ribulose 1:5 bisphosphate carboxylase, which converts ribulose 1:5 bisphosphate to 3-phosphoglycerate (Fig. 5.7) is activated allosterically by NADPH and ATP [31]. Thus CO_2 fixation can occur more rapidly if sufficient reductive power and energy are available for starch synthesis. 3-phosphoglycerate, the product of the reaction, acts as a *feed forward activator* of ADP glucose-pyrophosphorylase the rate limiting enzyme in the pathway of starch biosynthesis [32,33].

Light causes an influx of calcium ions into the chloroplast, which activates NAD kinase, the enzyme that converts NAD to NADP. Thus more NADPH is available for the reactions of the reductive pentose phosphate pathway. The regulation of NAD kinase by Ca^{2+} is mediated by calmodulin in an analogous manner to that described in Chapter 4 for cyclic AMP phosphodiesterase and adenylate cyclase [34]. There are substantial amounts of calmodulin in chloroplasts, and its structure and Ca^{2+}-binding properties are remarkably similar to those of calmodulin from mammalian tissues [35,36].

81

Thioredoxin, and a similar protein termed glutaredoxin, are present at high concentrations in all prokaryotic and eukaryotic cells that have been examined [37]. It is therefore possible that enzyme regulation by thiol–disulphide interchange is not confined to higher plants. However, reduced thioredoxin or glutaredoxin can also function as hydrogen donors for enzymes such as ribonucleotide reductase, which produces deoxyribonucleotides for DNA synthesis [38]. They may also simply function in the reduction of any protein disulphide bridges formed adventitiously within cells.

5.4 NAD$^+$ dependent ADP-ribosylation reactions

5.4.1 The mechanism of action of diphtheria toxin [39,40]

The disease diphtheria is caused by infection of the throat of man and certain other mammals with the bacterium *Corynebacterium diphtheriae*. However, while the bacteria themselves remain confined to the upper respiratory tract, many internal organs become damaged as a result of the infection. The reason for this is that *C. Diphtheriae* secretes a specific protein, which circulates throughout the body and is the causative agent of the disease. This toxin is lethal for laboratory guinea pigs within 4–5 days, at only 0.1 μg per kg body weight.

An understanding of the basis of this extreme toxicity began with the finding that the toxin completely blocked protein synthesis within an hour or two in cells grown in culture [41]. However, in cell free systems, inhibition was completely dependent upon NAD$^+$ as well as the toxin. The only component of the protein synthetic machinery affected was elongation factor 2 (EF-2), which is essential for the GTP dependent translocation of peptidyl tRNA from the acceptor to the donor site on the ribosomes, and for the movement of mRNA by one nucleotide triplet, after each round of peptide bond formation.

The discovery that diphtheria toxin was an enzyme which catalysed the ADP-ribosylation of EF-2 offered a simple explanation for its extreme toxicity.

$$\text{EF-2} + \text{NAD}^+ \;\rightleftharpoons\; \text{ADP-ribose-EF-2} + \text{Nicotinamide} + \text{H}^+$$

The ADP-ribosyl group is attached to one of the ring nitrogens of a derivative of histidine, termed diphthamide (Fig. 5.9) which has not been found in any protein other than EF-2 [42]. ADP-ribosylation of EF-2 does not affect its ability to interact with GTP or bind to ribosomes, but destroys its GTPase activity. Thus GTP is not hydrolysed to GDP and the translocation of peptidyl tRNA from the acceptor to the donor site on the ribosomes is blocked.

Diphtheria toxin as synthesized is *not* enzymically active. It must first be cleaved by a proteinase and its disulphide bridges reduced with a thiol, before activity is expressed. The mechanism of activation by trypsin is shown in Fig. 5.10. Trypsin 'nicks' the molecule at three closely spaced arginine residues (189, 191 and 192 from the amino-terminus). Reduction then yields an enzymically active N-terminal A-fragment, and an inactive

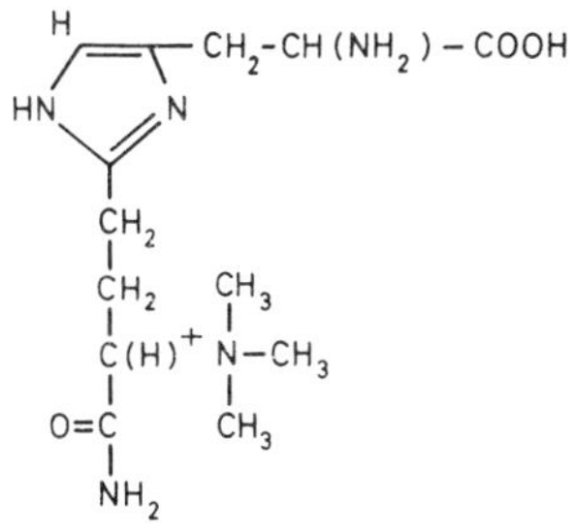

Fig. 5.9 Proposed structure of dipthamide, 2-[3-carboxyamido-3-(trimethylammonio) propyl]histadine [42].

C-terminal B-fragment. However neither fragment A, nor fragment B is toxic when added to cells in culture, in contrast to intact or nicked toxin.

Mutant toxins have been isolated that are non-toxic, and have defects in either the A-portion or the B-portion of the molecule. If a non-toxic mutant protein, defective in the A-portion is added to a non-toxic mutant protein defective in the B-portion in the presence of trypsin and a thiol, and the solution dialysed to remove the thiol and permit the reoxidation of disulphide bridges, toxicity is generated. These 'hybridization' experiments clearly show that the B-portion is essential for toxicity, and that toxicity is also dependent on the presence of an ADP-ribosylating activity. Nontoxic mutants in which the A-fragment is inactive, are competitive inhibitors of the native toxin in cell culture, suggesting that the toxin binds to specific receptor sites on the cell surface.

The intact or nicked toxin binds to cells through an interaction with a glycoprotein on the cell surface and a nucleotide binding site mostly located on the B-subunit [43]. This leads to entry of the A-subunit into the cell by a mechanism which still remains obscure. The toxin might be nicked *in vivo*, by proteinases present in blood (Chapter 3) or intracellularly. The disulphide bridges are presumably reduced intracellularly by glutathione or thioredoxin, allowing the A-fragment to catalyse the inactivation of EF-2. Protein synthesis is turned off in various internal organs, which in consequence become damaged. The process therefore involves two types of covalent modification, limited proteolysis and ADP-ribosylation.

A toxin from the unrelated bacterium *P. aeruginosa* catalyses exactly the same reaction as diphtheria toxin [44]. Since this toxin causes different

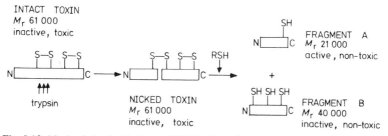

Fig. 5.10 Mechanism of activation of diphtheria toxin by trypsin.

83

clinical symptoms, such as hypotensive shock in dogs and monkeys, its cellular specificity must differ from diphtheria toxin. The stability of a toxin *in vivo* must also be a major factor in determining its toxicity.

5.4.2 The mechanisms of action of cholera, whooping cough and anthrax toxins

Cholera toxin is a protein secreted by the bacterium *Vibrio cholerae*, which irreversibly activates adenylate cyclase in all mammalian cells. It causes cholera by greatly increasing the level of cyclic AMP in intestinal cells. This results in the tremendous diarrhoea and serious water loss associated with the disease, since cyclic AMP plays an important role in controlling water absorption in the gut.

Cholera toxin contains one A-subunit (M_r 27 000) and five B-subunits (M_r 11 600), although the A-subunit is rapidly cleaved by proteinases into two fragments A1(M_r 22 000) and A2(M_r 5000) that remain linked by a disulphide bridge. The toxin binds to plasma membranes through the B-subunit, which interacts with ganglioside GM1, and this appears to initiate the uptake of the toxin into cells. The ability to activate adenylate cyclase is a property of the A1 fragment, which catalyses the ADP-ribosylation of the G_s protein (Chapter 4, Fig. 5.11). This appears to inhibit its GTPase activity, so that adenylate cyclase is permanently activated [43,45].

Several hormones produce inhibition, rather than activation of adenylate cyclase when they bind to their receptors; for example the interaction of adrenalin with its α_2 receptor, or enkephalins with the opiate receptor. These effects appear to be mediated by a further guanine nucleotide binding protein termed G_i (Fig. 5.11), which, like the G_s protein may also be a GTPase [46]. It is envisaged that inhibitory hormones stimulate the GTPase activity of the G_i protein, and so generate an inactive G_i–GDP

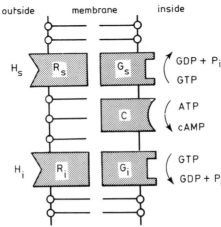

Fig. 5.11 Five component model for adenylate cyclase showing stimulatory and inhibitory pathways. R_s, stimulatory hormone receptor; R_i inhibitory hormone receptor; G_s, G_i, regulatory proteins mediating stimulation and inhibition of adenylate cyclase respectively; C, catalytic subunit of adenylate cyclase; H_s, H_i, stimulatory and inhibitory hormones [48].

complex. This would be in contrast to stimulatory hormones, which have been proposed to activate adenylate cyclase by facilitating the regeneration of G_s–GTP from G_s–GDP without affecting GTPase activity (Chapter 4). It has been found that one of the toxins secreted by the Gram-negative bacillus *Bordetella pertussis*, the pathogen responsible for whooping cough, catalyses the ADP-ribosylation of the G_i protein and abolishes α_2 adrenergic inhibition of adenylate cyclase, perhaps by inhibiting its GTPase activity (Fig. 5.11). This toxin has been called IAP (Islet activating protein) since it prevents the inhibition of insulin release by adrenalin in pancreatic islet β-cells [47].

The G_s protein (M_r 80 000) is composed of two subunits, M_r 45 000 and 35 000, and the G_i protein (M_r 80 000) is similarly composed of two subunits, M_r 41 000 and 35 000. The M_r 45 000 and M_r 41 000 subunits of G_s and G_i are ADP-ribosylated by cholera toxin and IAP respectively. The M_r 35 000 components of the two proteins are very similar and may even be identical [48].

Interestingly, another protein secreted by *Bordetella pertussis* is a soluble adenylate cyclase, which is essentially inactive in the absence of calmodulin [49]. It is believed that this protein may also enter mammalian cells and become activated by the calmodulin present within the cells that have been invaded. Thus infection with *Bordetella pertussis* may increase cellular cyclic AMP levels by two independent mechanisms.

The toxin responsible for anthrax is composed of three proteins termed protective antigen (PA), lethal factor (LF) and oedema factor (EF), which individually cause no known physiological effects, but are toxic in pairs. Injection of PA with LF causes death in rats within 60 min, whereas PA and EF cause oedema (swelling due to accumulation of water) in the skin of rabbits and guinea pigs. EF is an adenylate cyclase, which like the toxin of *Bordetella pertussis* is inactive in the absence of calmodulin. Current evidence suggests that PA interacts with cells to form a receptor system by which EF and LF gain access to the cytoplasm [50].

5.4.3 Inactivation of E. coli RNA polymerase by bacteriophage T4 infection

Corynebacterium diphtheriae only become toxigenic when they are themselves infected with a certain bacteriophage, and it is now clear that the structural gene for diphtheria toxin (and perhaps other toxins) originated on a phage genome [39]. It is therefore not surprising to learn that ADP-ribosylation is also used by bacteriophages to inactivate the functions of the bacteria they parasitize.

When bacteriophage T4 invades *E. coli* cells, the bacterium is switched from transcribing its own genome to the transcription of T4 genes within minutes; yet the enzyme used to transcribe the phage genes is the same *E. coli* RNA polymerase that transcribes *E. coli* genes. Two distinct ADP-ribosylations termed *alteration* and *modification* appear to play a role in this process [51].

Alteration is catalysed by an ADP-ribosyl transferase present in the 'head' region of T4, which enters the bacterium with T4 DNA. This leads to ADP-ribosylation of specific arginine residues on the polymerase (structure $\beta\beta'\alpha_2\sigma$). The α- and σ-subunits, and to a lesser extent, the β- and β'-subunits become altered. The stability of the linkages suggest an N-glycosidic linkage is formed between the ADP-ribosyl group and the guanidino group of arginine. *Alteration* takes place within a minute of infection, but no more than 50% of the α-subunits become altered. Moreover, *alteration* of the β-, β'- and σ-subunits reverses after a few minutes of infection. The precise role of *alteration* is unclear since phage multiplication can still occur in mutants lacking the ADP-ribosyl transferase [51].

Modification occurs several minutes after *alteration* and is dependent on protein synthesis within the infected cell. It leads to quantitative modification of a specific arginine residue on the α-subunit [52]. This may help to prevent the polymerase from transcribing the *E. coli* genome, but not T4 genes, suggesting that the α-subunits are essential for interaction of the polymerase with *promoter* sites on the *E. coli* genome [53]. *Alteration* and *modification* are catalysed by different enzymes, since alt$^-$ and mod$^-$ mutations map at different positions on the phage genome [54].

Summary

Although over 150 types of post-translational modifications of proteins have been identified *in vivo* [1], very few appear to be used for the control of enzyme activity. Thiol—disulphide interchange may be used to regulate many enzymes in photosynthetic organisms, and the presence of thioredoxin in mammalian cells [37] raises the possibility that this type of control is even more widespread. Regulation by reversible adenylylation, uridylylation and acetylation seems to be restricted to just a few bacterial enzymes. ADP-ribosylation of enzymes has only been demonstrated in cells that have been infected with bacterial toxins or viruses, although ADP-ribosylation of *non-enzymatic* proteins, such as histones, is well documented in the nuclei of mammalian cells [55].

The actions of toxins and viruses are strikingly similar to those of neural and hormonal stimuli, reflecting the fact that they must influence intracellular functions from extracellular locations. They have remarkable biological potency, their actions are initiated by interaction with receptors on the plasma membranes of target cells, and they frequently exert their effects through the covalent modification of intracellular proteins. In some cases, bacterial toxins and hormones even activate the same enzyme (adenylate cyclase). Many toxins are themselves enzymes, so that a single molecule could, in principle, catalyse the modification of all the available substrate within a cell. This explains their extreme toxicity.

References

[1] Wold, F. (1981), *Ann. Rev. Biochem.*, **50**, 783–814.

[2] Woolfolk, C. A. and Stadtman, E. R. (1964), *Biochem. Biophys. Res. Commun.*, **17**, 313–319; (1967), *Arch. Biochem. Biophys.*, **118**, 736–755.

[3] Stadtman, E. R., Shapiro, B. M., Kingdon, H. S., Woolfolk, C. A. and Hubbard, J. S. (1968), *Adv. Enz. Reg.*, **6**, 257–289.

[4] Shapiro, B. M. and Stadtman, E. R. (1970), *Ann. Rev. Microbiol.*, **24**, 501–524.

[5] Valentine, R. C., Shapiro, B. M. and Stadtman, E. R. (1968), *Biochemistry*, 7, 2143–2152.

[6] Holzer, H., Mecke, D., Wulff, K. Liess, K. and Heilmeyer, L. (1967), *Adv. Enz. Reg.*, 5, 211–225.

[7] Kingdon, H. S. and Stadtman, E. R. (1967), *Biochem. Biophys. Res. Commun.*, 27, 470–473.

[8] Shapiro, B. M., Kingdon, H. S. and Stadtman, E. R. (1967), *Proc. Nat. Acad. Sci.*, 58, 642–649.

[9] Kingdon, H. S., Shapiro, B. M. and Stadtman, E. R. (1967), in *Metabolic Interconversions of Enzymes 1966*, Springer-Verlag, Heidelberg, Vol. 58, pp. 1703–1710.

[10] Wulff, K., Mecke, D. and Holzer, H. (1967), *Biochem. Biophys. Res. Commun.*, 28, 740–745.

[11] Heinrikon, R. and Kingdon, H. S. (1971), *J. Biol. Chem.*, 246, 1099–1106.

[12] Stadtman, E. R., Ginsburg, A., Ciardi, J. E., Heh, J., Hennig, S. B. and Shapiro, B. M. (1970), *Adv. Enz. Reg.*, 8, 99–118.

[13] Ciardi, J. E., Cimino, F. and Stadtman, E. R. (1973), *Biochemistry*, 12, 4321–4330.

[14] Anderson, W. B. and Stadtman, E. R. (1970), *Biochem. Biophys. Res. Commun.*, 41, 704–709.

[15] Shapiro, B. M. (1969), *Biochemistry*, 8, 659–670.

[16] Anderson, W. B., Hennig, S. B., Ginsburg, A. and Stadtman, E. R. (1970), *Proc. Nat. Acad. Sci.*, 67, 1417–1424.

[17] Rhee, S. G., Chock, P. B. and Stadtman, E. R. (1978), *Proc. Natl. Acad. Sci.*, (USA), 75, 3138–3142.

[18] Stadtman, E. R., Brown, M., Segal, A., Anderson, W. A., Hennig, S. B., Ginsberg, A. and Mangum, J. H. (1972), in *Metabolic Interconversions of enzymes 1971*, (eds E. Helmreich, H. Holzer and O. Wieland) Springer-Verlag, Heidelberg, pp. 231–244.

[19] Brown, M. S., Segal, A. and Stadtman, E. R. (1971), *Proc. Nat. Acad. Sci.*, 68, 2949–2953.

[20] Mangum, J. H., Magni, G. and Stadtman, E. R. (1973), *Arch. Biochem. Biophys.*, 158, 514–525.

[21] Adler, S. P., Purich, D. and Stadtman, E. R. (1975), *J. Biol. Chem.*, 250, 6264–6272.

[22] Adler, S. P., Mangum, J. H., Magni, G. and Stadtman, E. R. (1974), in *Metabolic Interconversions of Enzymes 1973*, (eds E. H. Fischer, E. G. Krebs, H. Neurath and E. R. Stadtman), Springer-Verlag, Heidelberg, pp. 221–233.

[23] Garcia, E. and Rhee, S. G. (1983), *J. Biol. Chem.*, 258, 2246–2253.

[24] Ebner, E., Wolf, D., Gancedo, C., and Holzer, H. (1970), *Eur. J. Biochem.*, 14, 535–544.

[25] Segal, A., Brown, M. S. and Stadtman, E. R. (1974), *Arch. Biochem. Biophys.*, 161, 319–327.

[26] Schutt, H. and Holzer, H. (1972), *Eur. J. Biochem.*, 26, 68–72.

[27] Streicher, S., and Tyler, B. (1981), *Proc. Nat. Acad. Sci.*, (USA) 78, 229–233.

[28] Hubbard, J. S. and Stadtman, E. R. (1967), *J. Bacteriol.*, 94, 1007–1015.

[29] Gottschalk, G., Giffhorn, F. and Antranikian, G. (1982), *Biochem. Soc. Trans.*, 10, 324–326.

[30] Johnson, E. M. and Allfrey, V. G. (1978), in *Biochemical Actions of*

Hormones, **5**, (ed. G. Litwack) Academic Press, New York, pp. 1–51.

[31] Buchanan, B. B. (1980), *Ano. Rev. Plant. Physiol.*, **31**, 341–374.

[32] Preiss, J. (1978), *Advances in Enzymology*, **46**, 317–381.

[33] Copeland, L. and Preiss, J. (1981), *Plant Physiol.*, **68**, 996–1001.

[34] Anderson, J. M. and Cormier, M. J. (1978), *Biochem. Biophys. Res. Commun.*, **84**, 595–602.

[35] Anderson, J. M., Charbonneau, H., Jones, H. P., McCann, R. O. and Cormier, M. J. (1980), *Biochemistry*, **19**, 3113–3120.

[36] Jarrett, H. W., Brown, C. J., Black, C. C. and Cormier, M. J. (1982), *J. Biol. Chem.*, **257**, 13795–13804.

[37] Holmgren, A. (1981), *Trends in Biochemical Sciences*, **6**, 26–29.

[38] Thelander, L. and Reichard, P. (1979), *Ann. Rev. Biochem.*, **48**, 133–158.

[39] Collier, R. J. (1975), *Bacteriol. Rev.*, **39**, 54–85.

[40] Uchida, J. (1982), in *Molecular Aspects of Cellular Regulation*, 2, (eds S. Van Heyningen and P. Cohen), Elsevier Biomedical, Amsterdam, pp. 1–31.

[41] Strauss, N. and Hendee, E. D. (1959), *J. Exp. Med.*, **109**, 145–163.

[42] Van Ness, B. G., Howard, J. B. and Bodley, J. W. (1980), *J. Biol. Chem.*, **255**, 10710–10716.

[43] Van Heyningen, S. (1982), in *Molecular Aspects of Cellular Regulation*, 2, (eds S. Van Heyningen and P. Cohen), Elsevier Biomedical, Amsterdam, pp. 169–190.

[44] Iglewski, B. H. and Kabat, D. (1975), *Proc. Nat. Acad. Sci. (USA)*, **72**, 2284–2288.

[45] Ross, E. M. and Gilman, A. G. (1980), *Ann. Rev. Biochem.*, **49**, 533–564.

[46] Koski, G., Streaty, R. A. and Klee, W. A. (1982), *J. Biol. Chem.*, **257**, 14035–14040.

[47] Katadr, T. and Ui, M. (1982), *J. Biol. Chem.*, **257**, 7210–7216.

[48] Kahn, R. A., Katadr, T. Bokoch, G. M. Northup, J. K. and Gilman, A. G. (1983), in *Post-translational Covalent Modifications of proteins for function*, (ed. B. C. Johnson) Academic Press, New York (in press).

[49] Greenlee, D. V., Andreasen, T. J. and Storm, D. R. (1982), *Biochemistry*, **21**, 2759–2764.

[50] Leppla, S. H. (1982), *Proc. Nat. Acad. Sci. (USA)*, **79**, 3162–3166.

[51] Rohrer, H., Zillig, W. and Mailhammer, R. (1975), *Eur. J. Biochem.*, **60**, 227–238.

[52] Goff, C. G. (1974), in *Metabolic Interconversions of Enzymes 1973*, (eds E. H. Fischer, E. G. Krebs, H. Neurath and E. R. Stadtman), Springer-Verlag, Heidelberg, pp. 235–244.

[53] Mailhammer, R., Reiness, G., Ponta, H., Yang, H. L., Schweiger, M., Zillig, W. and Zubay, G. (1976), in *Metabolic Interconversions of Enzymes 1975*, (ed. S. Shaltiel) Springer Verlag Heidelberg., pp. 161–167.

[54] Rabussey, D. (1982), in *Mol. Asp. Cell Reg.*, 2, (eds S. Van Heyningen and P. Cohen), Elsevier Biomedical, Amsterdam, pp. 219–331.

[55] Okayama, H., Veda, K. and Hayaishi, O. (1978), *Proc. Nat. Acad. Sci. (USA)*, **75**, 1111–1115.

6 The nature of the allosteric transition

How does an allosteric effector, or a covalent modification such as phosphorylation, alter the activity of an enzyme, and what is the origin of the *sigmoid* saturation curves that are frequently obtained when activity is plotted as a function of substrate or effector concentration? In order to answer these questions it is first necessary to solve the three dimensional structure of an enzyme at high resolution, and then establish the changes in conformation that accompany the binding of substrates and effectors, or covalent modification. Of the enzymes discussed in Chapters 2, 4 and 5, only the structures of ATCase [1] and phosphorylase [2–4] have been solved to 0.3 nm resolution, but the conclusions about the nature of the allosteric transitions are still very tentative. In the case of ATCase there is some evidence that the transition from an active to an inactive state involves a small molecular expansion of the enzyme, and the preservation of the C:R−R:C unit (Fig. 2.12) implies that some change in the twist of the catalytic subunit *trimers* relative to the regulatory subunit *dimers* must take place.

Phosphorylase *b* is composed of two identical subunits (M_r 97 000) related symmetrically by a crystallographic two-fold axis. The binding site for the allosteric activator AMP (Fig. 6.1) is close to the subunit−subunit interface, and the nucleotide interacts with residues from both subunits. In particular, the adenine base is sandwiched between the side chains of Tyr-75 (from the AA' helix) and Val-45 from the CAP region of the symmetry related subunit. The 2' hydroxyl of the ribose hydrogen binds to the main chain carbonyl of Asn-44 from the symmetry related subunit. The phosphate of AMP is stabilized through an interaction with the guanidinyl group of Arg-309 from the BB' helix. The allosteric effector site is some 3.2 nm from the catalytic centre and the route of communication between these sites is not easy to see. However, it is possible that a movement of the BB' helix stimulated by interaction of the phosphate of AMP with Arg-309 could affect the loop carrying the residues Phe-285 and Phe-286 that are at the entrance to the catalytic site. Movement in this region could provide for subtle rearrangements of side chains at the active centre, leading to the generation of catalytic activity and improved affinity for substrate.

The first 19 residues from the N-terminus of phosphorylase *b* (containing Ser-14, the residue phosphorylated by phosphorylase kinase) have not been located by X-ray crystallography; and it has been concluded that this region must be mobile. In contrast, residues 8–19 are well ordered in phosphorylase *a* and the phosphate group on Ser-14 interacts with Arg-69 on the outside of the AA' helix and with the side chain of Arg-43 of the

89

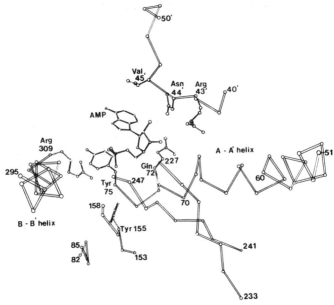

Fig. 6.1 The AMP binding site of glycogen phosphorylase *b* (Dr Louise Johnson, personal communication).

CAP region of the symmetry related subunit. The N-terminal tail carries four basic (Lys and Arg) groups in the vicinity of Ser-14, which, in the *b*-form, may serve to repel the tail from the recognition site for the seryl-phosphate (which contains a cluster of arginine residues).

The seryl-phosphate in phosphorylase *a* is some 1.5 nm from the AMP site and some 3.5 nm from the catalytic site. AMP and the seryl-phosphate both interact with residues on the AA' helix and with the CAP region. This suggests that there may be some common features in the mechanism by which AMP and phosphorylation cause the generation of catalytic activity. The details of this mechanism remain a mystery. The location of the phosphorylation site at the surface of the molecule was anticipated, since it must be accessible to phosphorylase kinase and protein phosphatase-1; and this situation must presumably apply to all interconvertable enzymes.

Many of our ideas about the origin of sigmoidal effects are derived from a consideration of the protein haemoglobin. Haemoglobin has an $\alpha_2\beta_2$ structure, where the two types of subunit each have a molecular weight near 16 000 and contain one haem group. The iron atom of each haem binds one molecule of O_2, so that a total of four are bound when haemoglobin is fully oxygenated. Oxy- and deoxyhaemoglobins have similar but distinct tertiary structures. A comparison of the two structures has suggested that when O_2 binds, the iron atom moves into the plane of the haem ring. Since the iron atom is rigidly attached to a particular histidine residue on the α and β-subunits, this triggers a series of discrete changes in the structure of each subunit, and a rearrangement of the positions of the subunits relative to one another [5,6].

It has been known for over 60 years, that when the oxygenation of haemoglobin is plotted against the partial pressure of O_2, a strongly sigmoidal curve is obtained. Inspection of the form of this curve suggests that its sigmoid character could arise if it was relatively difficult for the first molecule of O_2 to bind to any of the four haem groups, but that once this was accomplished, the binding of further O_2 molecules became progressively easier. Since the haem groups are too far apart to influence one another directly, and oxy- and deoxyhaemoglobins have different conformations, this suggested that the constraint on the binding of the first O_2 molecule might be related to the particular configuration of deoxyhaemoglobin.

This idea has been developed into two mathematical models which can theoretically account for sigmoid effects in proteins and enzymes. Readers are referred to the original papers [7,8] and to two excellent reviews [9,10] for the detailed descriptions of each model.

In the Koshland model [7], the binding of O_2 to the first subunit induces a structural change in that subunit, which is transmitted to an adjacent subunit, through the subunit contacts that link the two polypeptide chains together. The structural change induced in the second subunit causes it to bind O_2 more easily. This then affects the conformation of a third subunit and so on.

In the Monod model [8], haemoglobin is assumed to pre-exist as an equilibrium mixture of two conformational states; a 'tense' or T-state which binds O_2 with low affinity, and a relaxed or R-state which binds O_2 with high affinity. In the absence of O_2, haemoglobin is predominantly in the T-state. As O_2 is added, it binds preferentially to the R-state, shifting the equilibrium in the direction of R, and making available more high affinity binding sites for O_2. The combination of multiple binding sites for O_2, and the shift in the equilibrium between T and R, gives rise to the sigmoid effects.

The X-ray crystallographic analysis of the structure of haemoglobin has not been able to distinguish between these hypotheses. Crystallization is a process that freezes a molecule into one particular configuration, and analysis of the conformation of partially oxygenated molecules has proved impossible [6].

The Monod and Koshland models differ in that in the former, the protein is assumed to exist in two conformations in the absence of O_2 and the conversion of the T to the R state takes place by a *concerted* mechanism in which the conformation of all the subunits change simultaneously, while in the latter the conformational changes occur *sequentially* through a number of discrete steps. However, both models demand that the protein has the following structural features. It must exist in at least two conformations, there must be multiple binding sites for O_2 (and therefore multiple subunits), and the addition of oxygen must promote conversion to a form which binds oxygen with higher affinity.

It should be apparent from this brief qualitative account of the two theories, that the mechanisms which are thought to generate a sigmoidal response, are very much like an allosteric transition. The binding of O_2 at one site may influence the binding of oxygen at a remote site, by a mechanism analogous to the way in which CTP inhibits ATCase (Chapter 2) or cyclic AMP activates cAMP-PK (Chapter 4). Furthermore, the oxygenation curve of haemoglobin is perfectly analogous to the sigmoid curves that are sometimes obtained when enzyme activity is plotted as a function of substrate concentration (Figs. 2.4 and 2.11). The mechanism of action of many allosteric effectors can be explained very simply by these models. In the Monod terminology, an allosteric inhibitor would act by displacing the equilibrium toward the T-state, shifting the sigmoid saturation curve to the right making it more sigmoidal (e.g. the effect of alanine on pyruvate kinase – Chapter 4). Conversely, an allosteric activator would be promoting the formation of the R-state, moving the sigmoid saturation curve to the left, making it more hyperbolic (e.g. the effect of valine on threonine deaminase, Fig. 2.4, or F16P on pyruvate kinase, Chapter 4).

In practice, it can be difficult to distinguish between the *concerted* and *sequential* models experimentally, and it is even possible for both types of transition to operate in the same enzyme. For example, in the case of ATCase evidence has accumulated that the sigmoidal dependence of activity on aspartate concentration (Fig. 2.11) is best explained by a *concerted* transition, while the inhibition of activity by CTP (Fig. 2.11) is better explained by a sequential mechanism [1]. Thus the interactions between different catalytic subunits, and those between catalytic and regulatory subunits, occur through distinct molecular mechanisms.

Although both theories require that an enzyme be composed of two or more subunits, it is important to stress that *most* intracellular enzymes possess such a structure, yet very few show sigmoid kinetic behaviour or are regulated by allosteric effectors. This is conveniently illustrated by considering the enzymes that convert glycogen and glucose to lactic acid in mammalian muscle (Table 6.1). 10 of the 14 enzymes are formed from two or more subunits, yet only two show sigmoid kinetics or allosteric regulation. Moreover, one of the four *monomeric* enzymes that has no subunit structure (hexokinase) is regulated by an allosteric effector (G6P). G6P is a non-competitive inhibitor with respect to glucose, and a competitive inhibitor with respect to the second substrate, ATP [11]. There are no sigmoidal effects, and the allosteric transition must be mediated within a single subunit, as in the chymotrypsinogen–chymotrypsin conversion. A sigmoidal character is clearly not a prerequisite for the control of enzyme activity by an allosteric effector.

Besides the Koshland and Monod models, it is also possible to explain sigmoidal phenomena by a variety of kinetic models which do not require an enzyme to have multiple binding sites for the substrate [10]. It follows from these models that monomeric enzymes are also potentially capable of showing a sigmoidal dependence of activity on substrate concentration. In view of the remarkable variety of control mechanisms used by different enzymes, it seems entirely possible, that regulatory enzymes have evolved an equally varied set of mechanisms for generating sigmoidal effects, and

Table 6.1 Properties of the glycolytic enzymes from mammalian muscle.
(+)–activator; (−)–inhibitor; G3P, glyceraldehyde 3 phosphate.

Enzyme	Subunit $M.Wt. \times 10^{-3}$	Number of subunits	Allosteric regulation
Hexokinase	96	1	Yes (G6P⁻)
Phosphorylase	100	2	Yes (AMP⁺, ATP⁻, G6P⁻)
Debranching Enzyme	160	1	No
Phosphoglucomutase	62	1	No
Phosphohexose isomerase	65	2	No
Phosphofructokinase*	85	4	Yes (P_i^+, AMP⁺, F26P⁺, ATP⁻, citrate⁻)
Aldolase	40	4	No
Triose phosphate isomerase	26	2	No
G3P dehydrogenase	36	4	No
Phosphoglycerate kinase	45	1	No
Phosphoglycerate mutase	26	2	No
Enolase	41	4	No
Pyruvate kinase**	57	4	No
Lactate dehydrogenase	35	4	No

* See [12]

** Pyruvate kinase is a regulatory enzyme in liver (Chapter 4) but not in muscle.

that examples which illustrate all the theoretical possibilities will eventually be found.

It was pointed out in Chapter 2, that a sigmoidal response renders the activity of an enzyme more sensitive to fluctuations in the substrate or effector over a particular concentration range. However, virtually the only protein for which a sigmoidal effect has been demonstrated *in vivo* is haemoglobin, since the characteristic spectral changes that accompany oxygenation can easily be followed in intact red blood cells.

Concluding remarks

The major purpose of this book has been to illustrate the diversity of regulatory mechanisms in living cells that have been revealed by the study of purified enzymes, and to discuss the experimental problems involved in establishing their physiological significance. It has seemed of more biological significance, for example, to try and decide whether the activation of phosphorylase by AMP actually takes place within muscle cells, than to discuss which theoretical model accounts best for the sigmoidal dependence of activity on AMP concentration. A number of standard undergraduate textbooks in Biochemistry have, however, stressed these theoretical models to the virtual exclusion of other aspects of enzyme regulation. This study has, I hope, helped to redress this imbalance a little.

References

[1] Kantrowitz, E. R., Pastra-Landis, S. C. and Lipscomb, W. N. (1980), *Trends in Biochemical Sciences*, 5, 124–128.

[2] Weber, I. T., Johnson, L. N., Wilson, K. S., Yeates, D. G. R., Wild, D. L. and Jenkins, J. A. (1978), *Nature*, **274**, 433–437.

[3] Fletterick, R. J. and Madsen, N. B. (1980), *Ann. Rev. Biochem.*, **49**, 31–61.

[4] Jenkins, J. A., Johnson, L. N., Stuart, D. I., Stura, E. A., Wilson, K. A. and Zonotti, G. (1981), *Phil. Trans. R. Soc. London B.*, **293**, 23–41.

[5] Perutz, M. F. (1970), *Nature*, **228**, 726–734; **237**, 495–499.

[6] Benesch, R., Benesch, R. E. and Bauer, C. (1975), in *The Red Blood Cell*, (ed. D. Surgenor), Academic Press, London and New York, pp. 825–839.

[7] Koshland, D. E., Nemethy, G. and Filmer, D. (1966), *Biochemistry*, **5**, 365–385.

[8] Monod, J., Wyman, J. and Changeux, J. P. (1965), *J. Mol. Biol.*, **12**, 88–118.

[9] Koshland, D. W. (1969), *Curr. Top. Cell. Reg.*, **1**, 1–27.

[10] Newsholme, E. A. and Start, C. (1973), *Regulation in Metabolism*, Chapter 2, John Wiley and Sons, London.

[11] Colowick, S. P. (1973), in *The Enzymes* (3rd Edition), **IX**, (ed. P. D. Boyer) Academic Press, London, pp. 1–14.

[12] Hers, H. G. and Van Schaftingen, E. (1982), *Biochem. J.*, **206**, 1–12.

Index